Annals of Mathematics Studies

Number 63

TOPICS IN HARMONIC ANALYSIS

Related to the Littlewood-Paley Theory

BY

ELIAS M. STEIN

PRINCETON UNIVERSITY PRESS

AND THE

UNIVERSITY OF TOKYO PRESS

PRINCETON, NEW JERSEY

1970

L.C. Card: 72-83688
S.B.N.: 691-08067-4
A.M.S. 1968: 2265, 4201, 4750

Third printing, with corrections, 1985

4 6 8 10 9 7 5

Printed in the United States of America

PREFACE

This monograph contains essentially the material presented in a course given during the spring semester of 1968 at Princeton University. My purpose in these lectures was two-fold: First, to give a new approach to that part of harmonic analysis which for the sake of simplicity we refer to as the "Littlewood-Paley Theory." The techniques that are used lead to a wide generalization of results hitherto restricted to $\mathbf{R}^n$ and other special contexts.

My second aim was to give the interested student a rapid (although admittedly sketchy) introduction to various areas in analysis, in particular some elements of Lie groups, almost everywhere limit theorems in the context of martingales, and complex interpolation of operators. If I have succeeded in my two aims it is because the main tools used in Chapters III and IV come from martingale theory and interpolation theory, while interesting examples of the results may be obtained in the setting of compact and semi-simple groups.

I am deeply indebted to R. Gundy for several enlightening conversations and to C. Fefferman for his great care and effort in preparing the lecture notes.

NOTE TO THE THIRD PRINTING

In preparing this edition I have corrected several errors and added an appendix which outlines some more recent work. I wish to thank P. A. Meyer for his gracious help in carrying out this task.

CONTENTS

TOPICS IN HARMONIC ANALYSIS

Related to the Littlewood-Paley Theory

INTRODUCTION

Background

We shall use the phrase "Littlewood-Paley theory" rather loosely, to denote a variety of related results in classical harmonic analysis whose extension to a general setting is our main goal.

In its one-dimensional form the theory goes back to the 1930s and may be said to contain the Hardy-Littlewood maximal theorem, Hilbert transforms, the work of Littlewood-Paley in [16], and capped in effect with the multiplier theorem of Marcinkiewicz [35].[1]

This theory may be described in terms of the Poisson integral, which in $\mathbf{R}^1$ is given by the family of transformations

$$f(x) \to \frac{t}{\pi} \int_{-\infty}^{\infty} \frac{f(x-y)dy}{t^2+y^2} = u(x, t)$$

As is well known the behavior of the harmonic function $u(x, t)$ closely reflects the behavior of its boundary values. Now the main point of the so-called "complex method" is to pass to the analytic function $F(z)$ whose real part is u and to exploit complex function theory to study F and thus f. An example of this far-reaching idea arises in the Hilbert transform, which gives in effect the passage from the real to the imaginary parts of the boundary values of F.

The next stage of the development of this type of analysis which culminated in the 1950's, saw the primacy of complex function theory give way to real-variable methods, and this led to an extension of many of these

---

[1] See also Zygmund [20, Chapters 14 and 15].

results to R^n [2]. Characteristic of these techniques were various "covering lemmas" and certain singular integral transforms, whose kernels had a quite explicit description, all rather specific to R^n.

Main results

Our approach is essentially different from the above two and allows for a very general formulation of an essential core of the subject; it can be applied, in particular, to various new and interesting situations. Our starting point is the semi-group of Poisson integrals, that is we assume that we are dealing with a family of operators $\{T^t\}_{t\geq 0}$, defined simultaneously on $L_p(M)$ $1 \leq p \leq \infty$, with the property that

$$T^{t_1} \cdot T^{t_2} = T^{t_1+t_2}. \quad T^0 = I\ ,$$

and in addition to the usual measureability in t, it satisfies the following basic assumptions:

(I) T^t are *contractions* on $L_p(M)$, i.e., $\|T^t f\|_p \leq \|f\|_p$, $1 \leq p \leq \infty$.
(II) T^t are *symmetric*, i.e., each T^t is self-adjoint on $L_2(M)$.
(III) T^t are *positivity preserving*, i.e., $T^t f \geq 0$, if $f \geq 0$.
(IV) $T^t(1) = 1$.

We refer to the above as *symmetric diffusion semi-groups*.

The task we set ourselves is to develop, as far as is possible, the analogues of the Littlewood-Paley theory in the context of these semi-groups. The interest in this arises from the multiplicity of examples of symmetric diffusion semi-groups and the consequence this theory has for the eigenfunction expansions of their infinitesimal generators. Besides the usual Poisson integrals for R^n a variety of examples may be found in Chapter 2, Section 2 of Chapter 3, and in Chapter 5.

The main tools that are used are three-fold:

---

[2] See the bibliographic indications at the end of Chapter 2.

[3] It is to be noted that the assumptions (I)-(IV) are to some extent redundant.

(i) The spectral representation in $L_2(M)$ of T^t as $T^t = \int_0^\infty e^{-\lambda t}\, dE(\lambda)$, for an appropriate spectral family $E(\lambda)$. This is, of course, the direct substitute for the Fourier transform in R^n.

(ii) Connections of the semi-group T^t and certain auxiliary martingales and ergodic theorems.

(iii) Convexity properties of holomorphic families of operators which allow one to mediate between (i) and (ii).

A curious fact that should not be overlooked is that the theorems for martingales required in the technique (ii) were to a significant extent already anticipated by Paley in his paper [17] dealing with the Walsh-Paley series.

Among the results we obtain are:
The maximal theorem (in Chapter 3), to wit

$$\|\sup_{t>0} |T^t f(x)|\,\|_p \le A_p \|f\|_p, \qquad 1 < p \le \infty .$$

In Chapter 4 we prove the Littlewood-Paley type inequality

$$\left\| \left(\int_0^\infty t \left|\frac{\partial T^t}{\partial t} f\right|^2 dt \right)^{1/2} \right\|_p \le A_p \|f\|_p , \qquad 1 < p < \infty$$

and its converse.

We also obtain the multiplier theorem to the effect that if $T^t = \int_0^\infty e^{-\lambda t}\, dE(\lambda)$, and $T_m = \int_0^\infty m(\chi)\, dE(\chi)$, then

$$\|T_m(f)\|_p \le A_p \|f\|_p , \qquad 1 < p < \infty,$$

whenever

$$m(\lambda) = \lambda \int_0^\infty e^{-\lambda t} M(t)\, dt,$$

where M is bounded on $(0, \infty)$.

An interesting example arises if $m(\lambda) = \lambda^{i\gamma}$, for γ real.

The theorem for T_m may be viewed as a variant of the Marcinkiewicz multiplier theorem, but instead of requiring conditions on a finite number of of derivatives of $m(\lambda)$, it requires a specific analyticity.[4]

Another approach

Besides the general theory just described, to which Chapters 3 and 4 are devoted, still another approach is given, less general but on the other hand more fruitful in various instances. It arises whenever, roughly speaking, the square of the infinitesimal generator of T^t is a second-order elliptic operator, i.e., a "Laplacian." In that case the interplay between the x-derivatives and the t-derivative are also of interest, and it is possible to study the natural generalization of the Hilbert transform (the so-called Riesz transforms). This situation occurs in the context of compact Lie groups, and to some extent also in the non-compact Lie groups (in particular for the zonal functions of semi-simple groups); also for Sturm-Liouville expansions. This attack has as its starting point the universality of the identity

$$\Delta u^p = p(p-1)u^{p-2}|\nabla u|^2$$

where $u > 0$, $\Delta u = 0$, and Δ is an appropriate Laplacian with ∇ its associated gradient.

Organization

Since our purpose is also pedagogical we have included much illustrative material, in particular in Chapters II and V. Chapter I is intended merely as a brief review of some basic facts about Lie groups, to be read as an introduction for Chapters 2 and 5. The core of the monograph, Chapters II and III, however, stands independently of the other chapters.

[4] This is what is to be expected from a universal conclusion of this kind. Since it requires the same condition for $m(\lambda)$ for e.g., the case of R^n (all n), it would require that $m(\lambda)$ is C^∞. Other examples of symmetric diffusion semi-groups notably the one constructed in Section 2 of Chapter V, when $G = SL(2, R)$ shows that it is even *necessary* that m be analytic in λ.

CHAPTER I

LIE GROUPS (A REVIEW)

This chapter is a review consisting of various statements, often with only a bare indication of proof, which serve as the requisite background for what follows. In Sections 1 through 3 the main goal is the discussion of the Peter-Weyl theorem for compact groups. Sections 4 through 6 deal with Lie groups, their Lie algebras and the resulting universal enveloping algebra. Finally in Section 7 we describe the Laplacian for compact Lie groups which will be fundamental in Chapter II.

Section 1. Compact groups

We begin by summarizing some facts from the theory of integration of locally compact groups. For details see Weil [8, Chapter II] and Loomis [3, Chapter VI].

A *topological group* is a group G which is also a topological space, and whose group operations $\cdot : G \times G \to G$ and $^{-1}: G \to G$ are continuous. A *left Haar measure* on the locally compact topological group G is a Borel measure μ on G, satisfying the properties

(1) $\mu(K) < +\infty$ for K a compact subset of G;

(2) $\mu(O) > 0$ for $O \neq \phi$ an open subset of G. (We ignore the fact that O might not be Borel if G is too big.)

(3) $\mu(aE) = \mu(E)$ for any $a \in G$ and Borel set E.

The crucial property is, of course, (3), the "left-translation invariance" of μ. Here and elsewhere aE denotes $\{ax \mid x \in E\}$.

Every locally compact topological group has a left Haar measure, unique up to multiplication by a positive constant. For example, Lebesgue measure is a left Haar measure on the circle group S^1, being invariant under the left group-action, i.e., rotation; and every invariant measure on S^1 is a multiple of Lebesgue measure.

Haar measure on the circle group is not only left-invariant, but also right-invariant, i.e., $\mu(Ea) = \mu(E)$. A group with a left and right-invariant Haar measure is said to be *unimodular*. That the circle, or any abelian locally compact group, is unimodular, is quite trivial, for left and right-invariance mean the same thing. We can also see without difficulty that a compact topological group G is unimodular. For let μ be a left Haar measure on G, and pick any $a \in G$. The measure μ_a on G, defined by $\mu_a(E) = \mu(Ea)$ is obviously a left Haar measure. By the essential uniqueness of left Haar measure, $\mu_a = \Delta \cdot \mu$ for some positive constant Δ. But $\Delta = 1$ since $\mu(G) = \mu(G \cdot a) = \mu_a(G) = \Delta \cdot \mu(G)$. Therefore $\mu(Ea) = \mu(E)$ for all E, so that the left-invariant measure is also right-invariant. Not all groups are unimodular. A standard example is the group of all transformations $x \to ax + b$ from the real line to itself, where $a \neq 0$.

Suppose that a belongs to the unimodular locally compact group G, and that f is any function defined on G. The (left) *translate* of f by a is just the function $(\lambda(a)f)(x) \equiv f(a^{-1}x)$, $x \in G$; and the right translate is defined similarly, by $(\rho(a)f)(x) \equiv f(xa)$. Because of our choice of conventions, λ and ρ satisfy the composition laws $\lambda(a)(\lambda(b)f) = \lambda(ab)f$ and $\rho(a)(\rho(b)f) = \rho(ab)f$. Obviously $\lambda(1)f = \rho(1)f = f$.

Let us regard $\lambda(a)$ and $\rho(a)$ as operators taking functions to functions. Then $\lambda(a)$ and $\rho(a)$ are isometries of $L_p(G)$ where the L_p-spaces are defined with respect to the Haar measure on G, and $\lambda(a)^{-1} = \lambda(a^{-1})$. We shall prove the useful fact that $\lambda(a)f \to f$ and $\rho(a)f \to f$ in $L_p(G)$, as $a \to 1$ in G, for each fixed $f \in L_p(G)$. Since the operators $\lambda(a)$ and $\rho(a)$ are uniformly bounded, we need only consider $f \in C_0(G)$, the space of continuous functions with compact support. For $f \in C_0(G)$, however, $\lambda(a)f$ and

$\rho(a)f$ converge uniformly to f, and have their supports contained in a fixed compact set, provided that a varies in a small enough neighborhood of 1. So $\lambda(a)f \to f$ and $\rho(a)f \to f$ in L_p for $f \in C_0(G)$, which completes the proof. The result is, of course, valid only when $p < \infty$.

The *convolution* of two functions f and g on the group G, is defined by the formula $(f * g)(x) = \int_G f(y) g(y^{-1}x)dy$, where dy denotes Haar measure. If f and g belong to $L_1(G)$ then the integral defining $f * g(x)$ converges absolutely for almost every $x \in G$, so that $f * g$ is defined—in fact we have $\|f * g\|_{L_1(G)} \leq \|f\|_{L_1(G)} \|g\|_{L_1(G)}$. This follows easily from Fubini's theorem. Convolution is a well-behaved product operation. For instance $(f * g) * h = f * (g * h)$ and $(af + a'f') * g = a(f * g) + a'(f' * g)$. $f * g = g * f$ when the group G is commutative, but not in general. Nevertheless we can still guarantee $f * g = g * f$ on a non-commutative group, if f is a *central* (or *invariant*, or *class*) function, which means that $f(a^{-1}xa) = f(x)$ for $a, x \in G$. Equivalently, f is a central function if and only if $f(xy) = f(yx)$ for $x, y \in G$.

All of the above elementary observations follow from changes of variables and manipulations with multiple integrals. The reader may reconstruct the proofs or consult the cited references.

Henceforth, G denotes a compact topological group with its Haar measure dx normalized so that $\int_G dx = 1$.

We now turn to the representations of compact groups.

Let V be a finite-dimensional vector space over R or C. A *finite-dimensional representation* of G on V is a continuous homomorphism $g \to R_g$ which maps G into the group of non-singular linear transformations on V. Thus, a representation R of G on V satisfies

(1) $R_{g_1} R_{g_2}(v) = R_{g_1 g_2}(v)$ for $g_1, g_2 \in G$ and $v \in V$;

(2) $R_1 = I$, the identity mapping on V;

(3) $g \to R_g v$ is a continuous mapping of G into V for each fixed $v \in V$.

The representations R^1 on V^1, and R^2 on V^2, are called *equivalent* if there is an isomorphism of vector spaces $A: V^1 \to V^2$ (onto), such that $R^1_g = A^{-1}R^2_gA$ for each $g \in G$. Two equivalent representations are, in an obvious sense, different manifestations of the same object.

A *unitary representation* of G on the finite-dimensional Hilbert space H is a continuous homomorphism $g \to R_g$ which maps G into the group of unitary transformations of H. Two unitary representations R^1 on H^1, and R^2 on H^2, are *unitarily equivalent* if there is an isomorphism of Hilbert spaces $A: H^1 \to H^2$ (onto), such that $R^1_g = A^{-1}R^2_gA$ for each $g \in G$.

Every finite-dimensional representation of G is equivalent to a unitary representation. In other words, if R is a representation of G on a finite-dimensional vector space V, we may provide V with Hilber-space structure, so that the transformations R_g are all unitary. To prove this, we pick an arbitrary (strictly) positive-definite inner product $(\cdot, \cdot)_0$ on V. Define a new inner product by setting $(\phi,\psi) = \int_G(R_g\phi, R_g\psi)_0 dg$, for $\phi,\psi \in V$. The reader may easily verify that $(\cdot, \cdot)$ is an inner product and that $(R_g\phi, R_g\psi) = (\phi,\psi)$, so that R_g is, in fact, unitary.

The reason for the above lemma is that unitary representations are often easier to work with than ordinary representations. We shall see applications shortly.

Suppose that R and S are representations of G on vector spaces V and W respectively. The *direct sum* $R \oplus S$ is the representation on $V \oplus W$ given by $(R \oplus S)_g([v, w]) = [R_g v, S_g w]$, for $g \in G$ and $[v, w] \in V \oplus W$. The *tensor product* representation $R \oplus S$ on the vector space $V \otimes W$ is defined by $(R \otimes S)_g (v \otimes w) = (R_g v) \otimes (S_g w)$. Finally, the *contragredient* representation to R is the representation $g \to (R^t_g)^{-1}$ where "t" denotes the transpose. We leave to the reader the interpretation of these operations in terms of the matrices for the transformations R_g, S_g, etc. Note that if R is a unitary representation given by matrices $R_g = \{R_{ij}(g)\}$ then its contragredient is given by $R_g = \{\overline{R_{ij}}(g)\}$, where the term stands for complex conjugation.

An *invariant subspace* for a representation R of G on V, is a subspace $W \subseteq V$, such that $R_g(W) \subseteq W$ for each $g \in G$

A representation R on the space is *irreducible* if V has no non-trivial R-invariant subspaces, and it is *completely reducible* if it is (equivalent to) a direct sum of irreducible representations.

Every unitary representation (R, V) of G is completely reducible. The proof is by induction on dim V, which is called the *degree* of R. If R is equivalent to a non-trivial direct sum $S \oplus T$, then R is completely reducible, by an application of the inductive hypothesis to S and T. But if R does not split as a direct sum, then it is irreducible, hence completely reducible. For if W is any non-trivial R-invariant subspace of V, then the orthocomplement $W^{\perp}$ is also R-invariant. Therefore $R = R|_W \oplus R|_{W^{\perp}}$, contradicting the indecomposability.

We proved, using the compactness of G, that every finite-dimensional representation is unitary. So every finite-dimensional representation of G is completely reducible; which reduces the study of representations to that of irreducible representations.

Later on we will use the notion of *characters*. The character of a representation R of G, is the function $\chi_R(x) = \text{trace } R_x$ taking G into the complex numbers. χ_R depends only on the equivalence class of R, and $\chi_R(x) = \chi_R(a\,xa^{-1})$ for $x, a \in G$, i.e., χ_R is a central function. Eventually we shall show that χ_R determines R uniquely up to equivalence.

Let us illustrate some of the concepts we have defined, byconsidering the circle group $G = [0, 2\pi]$. The finite-dimensional irreducible representations are given by $\theta \to e^{ik\theta}$ for $\theta \in G$, $k \in Z$. The tensor product $e^{ik_1\theta} \otimes e^{ik_2\theta}$ is the representation $e^{i(k_1+k_2)\theta}$, and the contragredient representation to $e^{ik\theta}$ is $e^{-ik\theta}$. Thus, all the irreducible representations of G arise from the single representation $\theta \to e^{i\theta}$ by successive tensor products and contragredients.

(To prove that every irreducible representation is an $e^{ik\theta}$, argue as follows. An irreducible representation, of G is equivalent to an irreducible unitary representation R. of G on the space V. Now $\{R_g \mid g \in G\}$ is a commuting family of unitary operators, and can therefore be diagonalized simultaneously. Hence R is equivalent to a representation

$$\begin{pmatrix} S_{11}(g) & 0 & \\ & S_{22}(g) & \\ 0 & & \ddots \end{pmatrix}$$

But such a representation is the direct sum of the one-dimensional representations $S_{ii}(g)$. Therefore every irreducible representation of G is one-dimensional.) Observe that the irreducible representations $e^{ik\theta}$ of G are precisely the fundamental objects of Fourier analysis on G. In particular, the famous Parseval theorem states that we can decompose $L_2(G)$ as $\sum_{k=-\infty}^{\infty} \oplus \{e^{ik\theta}\}$, where $\{e^{ik\theta}\}$ is the one-dimensional space spanned by $e^{ik\theta}$, and the direct sum means the usual concept—that the $\{e^{ik\theta}\}$ are pairwise orthogonal, and that every $f \in L_2(G)$ is a countable linear combination of $e^{ik\theta}$. The appearance of representations in the analysis of the circle group is no coincidence. We shall prove that L_2 of any compact group splits as an infinite direct sum of finite-dimensional spaces arising from the irreducible representations of the group. This result is contained in the Peter-Weyl theorem, one of the goals of the present chapter.

As a first step to generalizing Fourier analysis on the circle, we shall prepare to prove an analogue of the orthonormality of $\{e^{ik\theta}\}$.

SCHUR'S LEMMA: (a) Let R^1 and R^2 be finite-dimensional representations of G, on vector spaces V^1 and V^2 which are irreducible. Suppose that the linear transformation $A: V^2 \to V^1$ satisfies $R_g^1 A = AR_g^2$ for all $g \in G$ (A is called an *intertwining* operator). Then either $A = 0$ or A is an isomorphism onto.

(b) Let R^1 be a finite-dimensional representation of G on a complex vector space V. If the transformation on A: $V \to V$ satisfies $AR^1_g = R^1_g A$ for all $x \in G$, then A is a multiple of the identity operator.

Proof: (a) The hypothesis $R^1_g A = AR^2_g$ shows immediately that ker A is an R^2-invariant subspace of V^2 and that im A is an R^1-invariant subspace of V^1. Since R^1 and R^2 are irreducible, ker A = 0 or V^2, and im A = 0 or V^1.

(b) Since $R_g(A - \lambda I) = (A - \lambda I)R_g$ for any $g \in G$, $\lambda \in C$, part (a) shows that $A - \lambda I$ is 0 or else is invertible. $A - \lambda I$ is not invertible if λ is an eigenvalue of A. QED.

THEOREM (Schur's Orthogonality Relations): Suppose that R^1 and R^2 are irreducible finite-dimensional unitary representations of G. Let $R^1_g = \{R^1_{ij}(g)\}$ and $R^2_g = \{R^2_{k\ell}(g)\}$. Then if R^1 and R^2 are inequivalent

(a) $$\int_G R^1_{ij}(x)\overline{R^2_{k\ell}(x)}dx = 0 \text{ for each } i, j, k, \ell;$$

(b) $$\int_G R^1_{ij}(x)\overline{R^1_{k\ell}(x)}dx = \delta^{k\ell}_{ij}/\text{degree}(R^1), \text{ where } \delta^{k\ell}_{ij} \text{ is}$$

the Dirac delta, set equal to one when $i = k$, $j = \ell$; 0 elsewhere.

Recall that Haar measure on G is normalized so that $\int_G dx = 1$.

Proof: (a) Let B be any linear transformation from V^2 into V^1, and set $A = \int_G R^1(x)BR^2(x^{-1})dx$. Nothing could be easier than to verify that $R^1_a A = AR^2_a$. Since R^1 and R^2 are not equivalent, A could not be an isomorphism. From Schur's Lemma we conclude that A = 0, By taking B a matrix with a single non-zero entry, we can easily show that

$$\int_G R^1_{ij}(x)\overline{R^2_{k\ell}(x)}dx = 0 .$$

(b) Let B be a linear transformation: $V^1 \to V^2$ and set $A = \int_G R^1(x)BR^1(x^{-1})dx$. As above, $R^1_a A = AR^1_a$. By Schur's lemma $A = \lambda I$, so $\int_G R^1(x)BR^1(x^{-1})dx = \lambda I$. Taking the trace of both sides, we find that $\mathrm{tr}(B) = \lambda \cdot \deg(R)^1$. If we take B to have a single 1 as its only non-zero entry, then trace (B) = 0 or 1 depending on the location of the non-zero entry. Recalling that $R^1_{k\ell}(x^{-1}) = \overline{R^1_{k\ell}(x)}$, we conclude as in part (a), that $\int_G R^1_{ij}(x)\overline{R^1_{k\ell}(x)}dx = \delta^{k\ell}_{ij}/\deg R^1$. QED

As advertised, the Schur orthogonality relations generalize the orthonormality of $\{e^{ik\theta}\}$ on the circle group.

As a consequence we see that if R^1 and R^2 are two irreducible unitary representations then

$$\int_G \chi_{R^1}(x)\overline{\chi_{R^2}(x)}dx = 1 \text{ or } 0$$

according to whether R^1 is or is not equivalent with R^2. In particular the irreducible representation R^1 is determined, up to equivalence, by its character.

Section 2. The Peter-Weyl Theorem

For a detailed discussion see Weil [8, Chapter V], Loomis [3, Chapter VIII], and Pontrijagin [6, Chapter IV].

In order to formulate the Peter-Weyl theorem, we need to generalize our notion of unitary representation. Suppose that H is any complex Hilbert space, finite or infinite dimensional. A map which associates to every $g \in G$, a linear operator R_g: $H \to H$, is a *unitary representation* of G on H if and only if

1. each R_g is unitary;
2. $g \to R_g$ is a homomorphism of G into the group of unitary operators on H;

3. For each fixed vector $\phi \in H$, the mapping $g \to R_g \phi$ is continuous from G to H (with the norm topology).

The notions of equivalence of arbitrary representations, direct sums any number of unitary representations, and of irreducibility of a unitary representation of G, are defined in the obvious way.

Our old friends $\lambda(a)$ and $\rho(a) : L_2(G) \to L_2(G)$ are unitary representations of G on $L_2(G)$. λ (respectively, ρ) is called the *left (right) regular representation* of G.

The space $\mathfrak{E}$ of *entry functions* is the linear space of functions on G spanned by the entries of finite-dimensional irreducible representations of G. Equivalently, $\mathfrak{E}$ is spanned by all functions of the form $x \to (R_x \phi, \psi)$, where (R, H) is an irreducible representation, and $\phi, \psi \in H$.

If G is the circle group, then $\mathfrak{E}$ is just the space of trigonometric polynomials.

We can now state the Peter-Weyl Theorem.

THEOREM:

(1) Every irreducible unitary representation of G is finite-dimensional.

(2) The right (or left) regular representation of G is the direct sum of finite-dimensional irreducible representations, and every equivalence class α of finite-dimensional irreducible representations appears in this direct sum as many times as d_α, the degree of that equivalence class.

(3) There are sufficiently many finite-dimensional representations to separate the points of G. That is, for any two different points $x, y \in G$, there is a suitable representation R such that $R_x \neq R_y$.

(4) $\mathfrak{E}$ is dense in $C(G)$ and in $L_2(G)$.

(5) Let Λ be the set of equivalence classes of finite-dimensional irreducible representations. For each $\alpha \in \Lambda$, pick a unitary representation R^α in the class α. If $f \in L_2(G)$, set $F(\alpha)$ equal to the

finite-dimensional matrix $\int_G f(x) R_x^\alpha \, dx$. ($F(\alpha)$ is analogous to a Fourier coefficient.)

Then $\|f\|_2^2 = \Sigma_{\alpha \epsilon \Lambda} d_\alpha \operatorname{tr}(F\alpha) F^*(\alpha)) = \Sigma_{\alpha \epsilon \Lambda} d_\alpha |||F(\alpha)|||^2$, $|||A|||$ denoting the Hilbert-Schmidt norm of the matrix $A = (a_{ij})$, given by $|||A|||^2 = \Sigma_{i,j} |a_{ij}|^2$.

We shall sketch a proof of the Peter-Weyl theorem.

For the time being, let us assume that G has a *faithful* f.d. representation. (A representation R_x is *faithful* if $x \neq y$ implies $R_x \neq R_y$.) This assumption greatly simplifies the proof. Later we will return to the general case. (Incidentally, although not all groups have such representation, most interesting groups are given as matrix groups already and obviously have faithful f.d. representations.) Begin with (4). Note that $\mathfrak{E}$ is an algebra. To prove this, we need only show that $f \cdot g \in \mathfrak{E}$ when $f(x) = R_{11}(x)$ and $g(x) = S_{11}(x)$, where R and S are f.d. irreducible representations of G. But obviously $f \cdot g$ is the (1, 1) entry in the representation $R \otimes S$. By decomposing $R \otimes S$ into irreducible representations, we may easily express $f \cdot g$ as a linear combination of entries of irr. rep. Thus $f \cdot g \in \mathfrak{E}$, so $\mathfrak{E}$ is an algebra. A similar argument, this time with the contragredient representation, shows that $\mathfrak{E}$ is closed under complex conjugation.

The trivial representation puts all constants into the algebra $\mathfrak{E}$. Finally, $\mathfrak{E}$ separates the points of G, by our assumption that G has a faithful f.d. representation. By the above, the Stone-Weierstrass theorem applies to $\mathfrak{E}$. Hence $\mathfrak{E}$ is dense in C(G), and therefore also in $L_2(G)$. This proves (4), and we noted (3) just a moment ago.

Let us turn to (2). Set $H = L_2(G)$, and suppose that α is an equivalence class of irreducible representations, of degree d_α. We define H_α as the subspace of H, spanned by the d_α^2 entries of a representation R(x) of class α. H_α is independent of the choice of R, and has dimension exactly d_α^2, by the Schur orthogonality relations.

Now $H_\alpha \perp H_\beta$ in H if $\alpha \neq \beta$, again by the Schur orthogonality relations. Furthermore, the space $\mathcal{E}$ is precisely the space of all finite linear combinations of vectors in the various H_α. From (4), we conclude that, in fact, $H = \Sigma_{\alpha \in \Lambda} \oplus H_\alpha$. This decomposition is *canonical*. To complete the proof of (2), we shall have to decompose H_α, and this cannot be done canonically.

For each α, pick a unitary representation (R^α, V^α) in the class α, and an orthonormal base $e_1 \cdots e_{d_\alpha}$ of V^α. Set $\phi_{ij}(x) = (R^\alpha_x e_i, e_j)$ the matrix entry. $\{\phi^\alpha_{ij}\}$ forms an orthogonal base for H_α. Since R^α is a representation of G, we obtain $\{\phi_{ij}(x)\}\{\phi_{ij}(a)\} = \{\phi_{ij}(xa)\}$ as matrices. In other words, $\Sigma_k \phi_{ik}(x)\phi_{kj}(a) = \phi_{ij}(xa)$. This equation shows that on the subspace $H_{\alpha i}$ of H_α, spanned by the vectors $\phi_{i1}, \phi_{i2}, \ldots, \phi_{id_\alpha}$, the right regular representation of G coincides with the representation R^α, i.e., $\rho(a)|_{H_{ai}} = R_a{}^\alpha$. This is a simple matter of disentangling notation.

So for each $i \leq d_\alpha$, we have a subspace $H_{\alpha i}$ on which the right regular representation coincides with R^α. Since $H = \Sigma_{\alpha \in \Lambda} \oplus H_\alpha = \Sigma_{\substack{\alpha \in \Lambda \\ i \leq d_\alpha}} \oplus H_{\alpha i}$, we have proved the decomposition of (2).

(5) is just the Parseval identity for the orthonormal base $\{d_\alpha^{+1/2}\phi^\alpha_{ij}(x)\}$ of H. ($\|d_\alpha^{+1/2}\phi^\alpha_{ij}(\cdot)\|_2 = 1$ follows from the Schur relations.)

Finally, (1) is left as a good exercise for the reader.

Thus, the Peter-Weyl theorem is proved, under the assumption that G has a faithful finite-dimensional representation.

Section 3. The Peter-Weyl Theorem (Concluded)

In this section we remove the restriction of a faithful f.d. representation from the proof of the Peter-Weyl theorem. To do so, we shall need some results from the elementary theory of Hilbert-Schmidt operators.

Let $(\mathfrak{M}, dx)$ be a measure space, $H = L_2(\mathfrak{M}, dx)$, and T be the operator on H defined by $(Tf)(x) = \int_{\mathfrak{M}} K(x, y) f(y) dy$, where K is some fixed function in $L_2(\mathfrak{M} \times \mathfrak{M})$. T is called a *Hilbert-Schmidt* operator. Every Hilbert-Schmidt operator is completely continuous, and the norm of a Hilbert-Schmidt operator, is dominated by its so-called *Hilbert-Schmidt norm* $(\int_{\mathfrak{M} \times \mathfrak{M}} |K(x, y)|^2 \, dx\,dy)^{1/2}$.

It is easy to check that a Hilber-Schmidt operator T, given by a kernel K is a self-adjoint operator on H if and only if $K(x,y) = \overline{K(y,x)}$ for almost all $x, y \in \mathfrak{M}$. If T is self-adjoint, then of course its eigenvalues are real, and if $\lambda_1 \neq \lambda_2$ then the eigenspaces $\{\phi \in H |\ T\phi = \lambda_1 \phi\}$ and $\{\phi \in H |\ T\phi = \lambda_2 \phi\}$ are orthogonal.

We shall use the following spectral theorem for Hilbert-Schmidt operators.

THEOREM. Let T be a self-adjoint Hilber-Schmidt operator given by a kernel $K(x, y)$, let $\lambda_1, \lambda_2, \ldots$ be the non-zero eigenvalues, counted according to their multiplicies (the multiplicity of an eigenvalue is the dimension of its eigenspace). Then $\Sigma \lambda_j^2 < \infty$: Let $\phi_1, \phi_2, \ldots$ be an orthonormal sequence of eigenvectors such that $T\phi_i = \lambda_i \phi_j$. Then $K(x, y) = \Sigma_j \lambda_j \phi_j(x) \overline{\phi_j(y)}$, where the infinite sum converges in $L_2(\mathfrak{M} \times \mathfrak{M})$.

See Riez-Nagy [34, Chapt. VI].

Note that for any appropriate T, we can find a sequence ϕ_i as above, by taking an orthonormal base for each eigenspace.

Now we can prove the Peter-Weyl theorem. First of all, we need only show that $\mathfrak{E}$ is dense in C(G), for that was the only place in our first proof of the theorem where we used the faithful representation.

To apply the theory of Hilbert-Schmidt operators, let K be any continuous complexvalued function on G, and define the ooerator T on $H = L_2(G)$, by $(Tf)(x) = \int_G f(y) K(y^{-1}x) dy$. Thus $Tf = f * K$. T is a Hilbert-Schmidt operator on H, and a bounded linear transformation from $L_2(G)$ to C(G). In addition, T is self-adjoint if $K(x) = \overline{K(x^{-1})}$.

The connection between the operator T and the Peter-Weyl theorem is that for any $\mu \neq 0$, the eigenspace $H_\mu \equiv \{\phi \in H |\ T\phi = \mu\phi\}$ is contained in

$\mathfrak{E}$. To prove this, observe that H_μ is finite-dimensional, since it an eigenspace of a completely continuous operator. Furthermore, H_μ is invariant under the left regular representation. For if $f * K = \mu f$, then $(\lambda(a)f) * K = \lambda(a)(f * K) = \lambda(a)(\mu f) = \mu(\lambda(a)f)$. Therefore, picking an orthonormal base $f_1 \cdots f_\ell$ of H_μ, we have for each $a \in G$, $f_j(a^{-1}x) = \Sigma_k R_{kj}(a)f_k(x)$, for a unique $R_{kj}(a)$. Obviously $\{R_{kj}(a)\}$ is a f.d. unitary representation of G. Setting $a = y^{-1}$ and $x = 1$ in the last equation, we have $f_j(y) = \Sigma_k R_{kj}(y^{-1})f_k(1)$. So, $f_j \in \mathfrak{E}$, which implies $H_\mu \subseteq \mathfrak{E}$.

Next we will show that if $\phi \in H$, then $T\phi \in C(G)$ may be uniformly approximated by an element of $\mathfrak{E}$. In fact, the spectral theorem for Hilbert-Schmidt operators, which we have quoted above, shows that we can write $\phi = \Sigma_{j=1}^\infty a_j \phi_j + \psi$ where $\Sigma a_j^2 < +\infty$ and $\psi \in \ker T$. (The ϕ_j are as in the spectral theorem.) For $\varepsilon > 0$ we may find N so lare that $\Sigma_{j>N} a_j^2 < \varepsilon^2$. Let us write $T\phi = T(\Sigma_{j=1}^N a_j\phi_j) + T(\Sigma_{j>N} a_j\phi_j) + 0$. Since T is bounded from $L_2(G)$ to $C(G)$, it follows that $\|T(\Sigma_{j\geq N} a_j\phi_j)\|_{C(G)} \leq C\varepsilon$, where C is the norm of $T: L_2(G) \to C(G)$. Hence

$$\left\| T\phi - \sum_{j=1}^N \lambda_j a_j \phi_j \right\|_{C(G)} = \left\| T\phi - T\left(\sum_{j=1}^N a_j\phi_j\right) \right\|_{C(G)} \leq \left\| T\left(\sum_{j>N} a_j\phi_j\right) \right\|_{C(G)}$$

$$\leq C\varepsilon .$$

Since $\Sigma_{j=1}^N (\lambda_j a_j)\phi_j \in \mathfrak{E}$ and $\varepsilon > 0$ is arbitrary, we conclude that $T\phi$ may be uniformly approximated by an element of $\mathfrak{E}$.

We have just proved that if $f \in L_2(G)$, $K \in C(G)$, and $K(x) = \overline{K(x^{-1})}$, then $f * K$ may be uniformly approximated by functions belonging to $\mathfrak{E}$. But given any f in $C(G)$, we have only to take K above with small support, and $\int K(x)dx = 1$, then $f * K$ uniformly approximates f. This completes the proof of the Peter-Weyl theorem.

The proof just given shows that $\mathfrak{E} \subseteq C(G)$, a fact implicit in statement (4) of the Peter-Weyl theorem. If G is a Lie group, we can also show that $\mathcal{E} \subseteq C^\infty(G)$. For, we have only to modify our proof that $\mathcal{E} \subseteq C(G)$ by using $K \in C^\infty(G)$ instead of merely $K \in C(G)$.

Closing Remarks.

The notation of this section is that of section 2.

A. The decomposition (1) of section 1 induces projections $E_\alpha: H \to H_\alpha$. We shall now show that E_α has an explicit form as a convolution operator; that, in fact, $E_\alpha(f)(x) = d_\alpha \int_G \chi_\alpha(y) f(y^{-1}x)dy$, where χ_α is the character of the representations in the class α. To verify this fact, it is enough to look at $f = \phi_{ij}^\beta$ since $\{d_\beta^{1/2}\, \phi_{ij}^\beta\}$ form an orthonormal base for H. Thus, we must show that

$$d_\alpha \int_G \chi_\alpha(y)\phi_{ij}^\beta(y^{-1}x)\,dy = \begin{cases} \phi_{ij}^\beta & \text{if } \beta = \alpha \\ 0 & \text{if } \beta \neq \alpha \end{cases}.$$

By definition of $\{\phi_{ij}^\alpha\}$ and χ_α, $\chi_\alpha(x) = \mathrm{tr}(\{\phi_{ij}^\alpha(x)\}) = \Sigma_\ell\, \phi_{\ell\ell}^\alpha(x)$. Since $\{\phi_{ij}^\beta\}$ is a unitary representation of G, $\phi_{ij}^\beta(y^{-1}x) = \Sigma_k \phi_{ik}^\beta(y^{-1})\phi_{kj}^\beta(x)$ $= \Sigma_k\, \overline{\phi_{ki}^\beta(y)}\, \phi_{kj}^\beta(x)$. So

$$d_\alpha \int_G \chi_\alpha(y)\phi_{ij}^\beta(y^{-1}x)\,dy$$

$$= d_\alpha \sum_{k,\ell} \phi_{kj}^\beta(x) \int_G \phi_{\ell\ell}^\alpha(y)\overline{\phi_{ki}(y)}\,dy = \begin{cases} \phi_{kj}^\beta & \text{if } \alpha = \beta \\ 0 & \text{if } \alpha \neq \beta \end{cases},$$

by Schur's orthogonality relations. The verification is complete.

B. By part (5) of the Peter-Weyl theorem, there is one-to-one correspondence between $f \in L_2$ and the corresponding sequence of "Fourier coefficients" $\{F(\alpha)\}_{\alpha \in \Lambda}$. It may easily be shown that if $f \sim \{F(\alpha)\}_{\alpha \in \Lambda}$, then $f(a^{-1}x) \sim \{R_x^\alpha(a)F(\alpha)\}_{\alpha \in \Lambda}$ and $f(xa) \sim \{F(\alpha)R_x^\alpha(\alpha)\}_{\alpha \in \Lambda}$, and that if $f \sim \{F(\alpha)\}$ and $g \sim \{G(\alpha)\}$ then $f * g \sim \{F(\alpha)\,G(\alpha)\}$.

C. The basic operators of Fourier analysis are those which commute with translations. These operators are characterized in the following result.

THEOREM: Let T be a bounded operator from $L_2(G)$ to $L_2(G)$.
(a) T commutes with the right regular representation if and only if there is a collection $\{M_\alpha\}_{\alpha \in \Lambda}$ of matrices such that $g = Tf$ with $g \sim \{G(\alpha)\}$, $f \sim \{F(\alpha)\}$ means precisely that $G(\alpha) = M_\alpha F(\alpha)$. (N.B. M_α acts on the left.) The norm of T is $\sup_\alpha \|M_\alpha\|$. A similar result characterizes left-translation invariant operators.

(b) T commutes with both the left and right regular representation if and only if all the above matrices M_α are constant multiples of the identity matrix.

Let $T\phi_{ij}^\alpha = \psi_{ij}^\alpha(x) \in L_2(G)$. Since $\phi_{ij}^\alpha(xa) = \Sigma_k \phi_{ik}^\alpha(x)\phi_{kj}^\alpha(a)$ and T commutes with the right regular representation,

$$\psi_{ij}^\alpha(xa) = T(\phi_{ij}^\alpha(xa)) = \sum_k \psi_{ik}^\alpha(x)\phi_{kj}^\alpha(a) .$$

Setting $x = 1$, $a = y$, we now have

$$\psi_{ij}^\alpha(y) = \sum_k \psi_{ik}^\alpha(1)\phi_{kj}^\alpha(y) .$$

Part (a) now follows easily, with $M_\alpha = \{\psi_{ik}^\alpha(1)\}$. Part (b) is immediate from part (a) and Schur's lemma.

D. Using the preceding theorem we may characterize the central functions on G. For if f is a central function in $L_1(G)$, then $g \to f * g$ commutes with the left and right regular representations. From our theorem, we deduce that $f \sim \{F(\alpha)\}_{\alpha \in \Lambda}$ where each $F(\alpha)$ is a multiple of the identity matrix.

What does it mean that $F(\alpha)$ is a constant multiple of I, the identity?

If, for example, $F(\alpha) = \begin{cases} 0 & \text{if } \alpha \neq \beta \\ I & \text{if } \alpha = \beta \end{cases}$, then f ia just the character of the representation β. So in general, a central function is, at least formally, a countable linear combination of characters on G.

We list two concrete applications of this heuristic principle. Every continuous central function f on G is a uniform limit of linear combinations of characters on G.

Every L_2 central function f on G is the L_2 limit of linear combinations of characters on G.

These results can be proved by noting that f may be approximated in the appropriate norm by a finite linear combination $\Sigma\, A^{\alpha}_{ij}\, \phi^{\alpha}_{ij}(x)$ (see statement (4) of the Peter-Weyl theorem). Now $f(x) = \int_{a\in G} f(axa^{-1})\,da$, while $\int_{a\in G} \phi^{\alpha}_{ij}(axa^{-1})\,da = d_{\alpha}^{-1}\delta_{ij}\,\chi_{\alpha}(x)$. Details are left to the reader.

These observations show that although the theory of compact topological groups is not essentially commutative, there is an important commutative part of the theory, namely the study of central functions and characters. Fourier analysis will be especially concerned with this part of the theory.

Exercise for the reader: Show, using Schur orthogonality, that a f.d. representation is uniquely determined by its character.

Section 4. Lie groups; examples.

The purpose of the next three sections is to sketch some portions of the theory of Lie groups which we shall need later on. Useful references are:

1. Nomizu, *Lie Groups & Differential Geometry,* [4].
2. Chevalley, *Theory of Lie Groups,* [1].
3. Pontrjagin, *Topological Groups* (1st and 2nd editions), [6] and [7].
4. Helgason, *Differential Geometry and Symmetric Spaces,* [2].

A *Lie group* is a group G, which is also a C^∞-manifold, such that the group operations $(a, b) \in G \times G \to ab \in G$ and $a \to a^{-1}$ are C^∞-functions. It can be shown that every Lie group has a real-analytic structure which makes the group operations real-analytic.

Two Lie groups $G_1 \times G_2$ are *isomorphic* ($G_1 \cong G_2$) if there is a group isomorphism $j: G_1 \to G_2$ (onto) which is also a diffeomorphism. G_1 and G_2 are *locally isomorphic* ($G_1 \sim G_2$) if there are neighborhoods $N_1 \subseteq G_1$ and $N_2 \subseteq G_2$ of the identity and a diffeomorphism $j: N_1 \to N_2$ (onto) such that

a. If $x, y, x \cdot y \in N_1$ then $j(x)j(y) = j(xy)$;

b. If $x, y, x \cdot y \in N_2$ then $j^{-1}(x)j^{-1}(y) = j^{-1}(xy)$.

For example, the real line R^1 is locally isomorphic to the circle group.

We shall use the notion of local isomorphism to classify compact connected Lie groups. First we give the easy part of the classification.

THEOREM. In each equivalence class of locally isomorphic connected Lie groups, there is a unique simply-connected group $\tilde{G}$. Every group G in the equivalence class is of the form $G = \tilde{G}/Z$, where Z is a discrete central subgroup of $\tilde{G}$. Conversely, $G = \tilde{G}/Z$ is locally isomorphic to $\tilde{G}$ if Z is a discrete central subgroup.

$\tilde{G}$ is called the *universal covering group* of G.

The theorem is proved by picking any group G of the equivalence class and setting $\tilde{G}$ equal to the universal covering space of G. It is very easy to impose a group structure on $\tilde{G}$. Details may be found in Chevalley [1]. Notice that the fact that Z is central is immediate from the observation that the fibre Z is a discrete normal subgroup of $\tilde{G}$. For if $z \in Z$ is arbitrary, aza^{-1} will be close to z if a belongs to a small enough neighborhood of 1 in $\tilde{G}$. On the other hand $aza^{-1} \in Z$, which implies that $aza^{-1} = z$ for a close to 1, since Z is discrete. $\tilde{G}$ is connected and is therefore generated by any neighborhood of 1. Hence $aza^{-1} = z$ for *any* $a \in \tilde{G}$. Thus Z is central.

Having determined the structure of each equivalence class of connected Lie groups, we are left with the immensely more difficult task of classifying (connected) Lie groups up to local isomorphism. For *compact* groups, the solution is given in terms of the following:

1. The circle group $T^1 = R^1/Z$; more generally, the n-*torus* $T^n = T^1 \times \cdots \times T^1$ (n factors). These are the only compact connected abelian Lie groups (see Chevalley [1], p. 212-213).
2. The group SO(n) $(n \geq 3)$ of all orthogonal n-dimensional matrices of determinant + 1. SO(n) is called the *special orthogonal group.* When $n = 2k$, SO(n) is called a D_k *group;* when $n = 2k + 1$, SO(n) is called a B_k *group.*
3. The *special unitary group* SU(n) $(n \geq 2)$, the group of all unitary n-dimensional complex matrices of determinant 1. This gives the A -*series* of groups, $A_{n-1} = SU(n)$.
4. The *symplectic group* Sp(n), the quaternionic analogue of the real group SO(n), and its complex version SU(n). More explicitly, let Q^n denote quaternionic n-space, a vector space over the quaternion field Q. Recall that for each quaternion $a = \alpha_1 + \alpha_2 i + \alpha_3 j + \alpha_4 k$, the *quaternion conjugate* $\bar{a}$ is defined as $\alpha_1 - \alpha_2 i - \alpha_3 j - \alpha_4 k$. Using the quaternion conjugate, we can define the "inner product" on Q^n by $(a, b) = \sum_{k=1}^{n} a_\ell \bar{b}_\ell$ for $a = (a_1 \cdots a_n)$, $b = (b_1 \cdots b_n) \in Q^n$. Sp(n) consists exactly of those transformations of Q^n, linear over Q, and preserving the inner product. Sp(n) is also designated C_n.
5. The *exceptional groups* E_6, E_7, E_8, F_4, G_2, which we cannot describe here.

The classification of compact connected groups is given by the following result.

THEOREM. Every compact connected Lie group is locally isomorphic to a finite product of the groups listed above.

The proof is too difficult and long to be included here.

A few remarks are in order. First of all, the product of basic groups, mentioned in the theorem is unique, except for the following redundancies:

1. $SO(3) \sim SU(2) \cong Sp(1)$
2. $SO(4) \sim SO(3) \times SO(3) \sim SU(2) \times SU(2)$
3. $SO(5) \sim Sp(2)$
4. $SO(6) \sim SU(4)$

Secondly, it is possible to show that except for T^n, all the groups in our list have compact universal covering groups. In particular, Spin (n), the *spinor group,* defined as the universal covering group of $SO(n)$, is compact. It can be shown that $SO(n) = Spin(n)/Z_2$.

Finally, the reader is entitled to an explanation of the cryptic symbols $A_k, B_k, C_k, D_k, E_6, E_7, E_8, F_4, G_2$. The notation is based on the concept of *rank.* The rank of a compact Lie group is the largest integer k such that the group has a k-torus embedded in it. Equivalently, the rank of a group G is the highest dimension of any abelian subgroup of G. The groups A_k, B_k, C_k, and D_k all have rank k, E_6 has rank six, and so forth.

The importance of the classification theorem for us, is that it gives us a bird's eye view of what Fourier Analysis on compact Lie groups might be like. For among the basic groups 1–5, T^n is the setting for classical Fourier series, and on $SO(n)$ much of the classical theory has been carried over.

Section 5. Lie algebras

We shall (eventually) introduce the *Lie Algebra* of a Lie group G. Recall first that if M is an n-dimensional C^∞-manifold and p is any point of M, the tangent space at p, $T_p(M)$ is the n-dimensional vector space of all linear functionals L on $C^\infty(M)$ which satisfy $L(fg) = L(f)g(p) + f(p)L(g)$.

$$L(fg) = L(f)g(p) + f(p)L(g) \ .$$

A *vector field* is a linear mapping $X: C^\infty(M) \to C^\infty(M)$ which satisfies the condition $X(fg) = (Xf)g + f(Xg)$. This is equivalent to the usual definition. The *bracket* operation assigns to any two vector fields X and Y, a vector field $[X, Y]$ defined by $[X, Y](f) = X(Yf) - Y(Xf)$. The bracket operation is bilinear, anti-symmetric, and satisfies Jacobi's identity

$$[X, [Y, Z]] + [Z, [X, Y]] + [Y, [Z, X]] = 0.$$

The space of all vector fields is infinite-dimensional, and so too big for our purposes. We therefore restrict our attention to the space of all *left invariant* vector fields. The vector field X on G is called left invariant if for any $a \in G$, $\lambda(a)X = X\lambda(a)$, where λ denotes the left regular representation of G. In other words, a left-invariant vector field commutes with left translations. The space of left-invariant vector fields is closed under the bracket operation.

Every vector field X on a Lie group G determines an element X_1 of the tangent space to G at 1, defined by $X_1(f) = (X(f))(1)$ for $f \in C^\infty(G)$. It is easy to show that $X \to X_1$ is an isomorphism of the space of left-invariant vector fields on G, onto the tangent space $T_1(G)$. This isomorphism induces a bracket product $[\cdot, \cdot]$ on $T_1(G)$. $T_1(G)$ with its bracket product, is called the *Lie algebra* $\mathfrak{g}$ of the Lie group G.

The process which defines the Lie algebra is essentially differentiation. In particular, a vector field, being a section of the tangent bundle is really nothing but a first-order differential operator on G. We shall define a process, analogous to integration, which takes us from the Lie algebra back to the Lie group.

A family $\{\phi_t\}$ is called a *one-parameter group of diffeomorphisms* of the n-manifold M if

1. ϕ_t is a diffeomorphism of M onto itself, for each t, $-\infty < t < +\infty$.
2. The map $(t, p) \to \phi_t(p)$ taking $R^1 \times M$ into M is smooth.
3. $\phi_t \circ \phi_s = \phi_{t+s}$ and ϕ_0 is the identity.

We can associate a vector field X to each one-parameter group $\{\phi_t\}$, by setting

$$(Xf)(p) = \lim_{t \to 0} \frac{f(\phi_t(p)) - f(p)}{t} = \frac{d}{dt} f \circ \phi_t(p)\Big|_{t=0},$$

for $f \in C^\infty(M)$, $p \in M$.

The converse problem is not, in general, solvable. That is, given a vector field X, there may not be any one-parameter group $\{\phi_t\}$ which satisfies $(Xf)(p) = \frac{d}{dt} f \circ \phi_t(p)|_{t=0}$. But there does exist a *local* one-parameter group for which it holds. More precisely, given any point $p \in M$, there is a neighborhood $N \subseteq M$ of p, an $\varepsilon > 0$, and a family $\{\phi_t\}_{|t| \leq \varepsilon}$ of mappings, defined only on N, such that

1′. ϕ_t is a diffeomorphism of N into M, for each t ($|t| < \varepsilon$).

2′. $(t, q) \to \phi_t(q)$ is smooth.

3′. If $|t_1| < \varepsilon$, $|t_2| < \varepsilon$, $|t_1 + t_2| < \varepsilon$ and $q \in N$, $\phi_{t_2}(q) \in N$; then $\phi_{t_1}(\phi_{t_2}(q)) = \phi_{t_1+t_2}(q)$. $\phi_0(q) \equiv q$.

4′. $(Xf)(q) = \frac{d}{dt}(f \circ \phi_t(q))\Big|_{t=0}$ for $q \in N$, $f \in C^\infty(M)$.

Moreover, X (essentially) uniquely determines $\{\phi_t\}$.

The proof is an easy application of the Cauchy-Lipschitz existence theorem of ordinary differential equations. For complete details, see Nomizu [4], Chapter I.

Returning to Lie groups, let X be an element of the Lie algebra of the connected Lie group G. We may think of X as a left-invariant vector field on G. By the above, there is an essentially unique local one-parameter group $\{\phi_t\}$ satisfying 1′ to 4′ above. Using the uniqueness of $\{\phi_t\}$ and the left-invariance of X, we can easily show that ϕ_t commutes with left translations, i.e., $a \cdot \phi_t(x) = \phi_t(ax)$ for $a, x \in G$, whenever this equation makes sense. Setting $x = 1$ we find that $\phi_t(a) = a \cdot \phi_t(1)$. Thus ϕ_t is nothing but a multiplication *on the right* by an element $a(t) \in G$. The group property of $\{\phi_t\}$ say that $a(t_1)a(t_2) = a(t_1 + t_2)$. (Also $a(0) = 1$.) So far $a(t)$ is only defined for t in a neighborhood of 0, but we can easily extend a to a homomorphism $a: t \to a(t)$ from the real line into G.

We have proved: For any left-invariant vector field X on G, there is a unique homomorphism $a: R \to G$ such that

$$(Xf)(x) = \frac{d}{dt} f(x \cdot a(t))\Big|_{t=0} .$$

We denote $a(t)$ by $\exp(tX)$.

The *exponential map* exp from the Lie algebra $\mathfrak{g}$ to the Lie group G is now easy to define. We simply set $\exp(X) = \exp(1 \cdot X)$.

The exponential is a powerful tool in the study of Lie groups, for it tells very precisely, the relations between a Lie group and its Lie algebra. Thus, one can show using the exponential, that two groups are locally isomorphic if and only if their Lie algebras are isomorphic.

A calculation of Jacobians shows that $\exp: \mathfrak{g} \to G$ is a local diffeomorphism. Hence any base $X_1 \cdots X_n$ of $\mathfrak{g}$ induces a system of local coordinates at $1 \in G$, given by $(t_1, \ldots, t_n) \to \exp(t_1X_1 + \cdots + t_nX_n) \in G$; the system is called the *canonical* co-ordinate system with respect to the base $X_1 \cdots X_n$.

The following formulas are useful for the application of canonical coordinates.

1. (Taylor's formula) If $f: G \to R$ is real-analytic, and $x \in G$, then

$$f(x \cdot \exp tX) = \sum_{n=0}^{\infty} \frac{t^n}{n!} (X^n f)(x), \quad \text{for } X \text{ in a neighborhood of zero in } \mathfrak{g}.$$

2. For $X, Y \in \mathfrak{g}$, $\exp tX \cdot \exp Y =$

$$= \exp(t(X+Y) + \frac{t^2}{2}[X, Y] + O(t^3)) \quad \text{as } t \to 0.$$

3. For $X, Y \in \mathfrak{g}$, $(\exp tX)(\exp tY)(\exp tX)^{-1}(\exp tY)^{-1}$

$$= \exp(t^2[X, Y] + O(t^3)) \quad \text{as } t \to 0.$$

Formula 1. follows if we apply Taylor's theorem to the auxilliary function $F(t) = f(x \cdot \exp tX)$, keeping in mind that $(d/dt)f(x \cdot \exp tX)|_{t=0} = (Xf)(x)$ by definition of the exponential.

To prove formula 2., we select any real-analytic function f defined on G. Using formula 1., we can easily check that the Taylor expansions of $t \to f(\exp tX \cdot \exp tY)$ and $t \to f(\exp(t(X+Y) + \frac{t^2}{2}[X,Y]))$ agree up to second order. Since f is essentially arbitrary, we conclude that $\exp tX \cdot \exp tY$ and $\exp(t(X+Y) + \frac{t^2}{2}[X, Y])$ differ by a third-order quantity, in the obvious sense. Formula 2. now follows from the fact that the exponential map is a local diffeomorphism.

Finally, formula 3. is a trivial consequence of formula 2.

Details of these hastily sketched proofs are left to the reader or to the cited literature.

Let us now consider some examples. First of all, let $G = GL(n, R)$, the general linear group. Although G has two connected components, the Lie algebra and exponential map still make sense for G. Since G is an open subset of the vector space $M(n, R)$ of all real $n \times n$ matrices, we can identify the tangent space at $I \in G$ with $M(n, R)$. Thus, the Lie algebra of G is canonically isomorphic to $M(n, R)$, as a vector space. To comthe exponential map, we shall find all one-parameter groups in G. If $t \to N(t)$ is a one-parameter group, then of course $N(s + t) = N(s)N(t)$. Differentiating this equation in s, we obtain

$$\frac{dN}{dt}(t) = AN(t), \text{ where } A = \frac{dN}{dt}(t)\Big|_{t=0} \in M(n,R).$$

This differential equation, together with the initial condition $N(0) = I$, has only one solution.

$$N(t) = e^{tA} \equiv \sum_{n=0}^{\infty} \frac{t^n}{n!} A^n .$$

Therefore the one-parameter subgroups of G are $t \to e^{tA}$, $A \in M(n, R)$.

An easy computation shows that the tangent vector (i.e., element of the Lie algebra) induced by e^{tA} is just A. So the exponential map must be given by $\exp(tA) = e^{tA}$, i e., $\exp(A) = e^{A}$, which justifies the name "exponential." Formula 2. above, now shows that the bracket operation on the Lie algebra $M(n, R)$ is just $[A, B] = AB - BA$.

Our next example is $G = SO(n)$, the special orthogonal group. Since G is a subgroup of $Gl(n, R)$, it follows from the first example, that the Lie algebra is a subalgebra of $M(n, R)$ (given the natural bracket product $[A, B] = AB - BA$). In particular, the one-parameter subgroups of $SO(n)$ are all of the form $t \to e^{tA}$. But from elementary linear algebra we know that e^{A} is orthogonal if and only if A is skew-symmetric, i.e., $(Ax, y) = -(x, Ay)$, So the Lie algebra of $SO(n)$ consists of all real skew-symmetric $n \times n$ matrices, under the natural bracket product; and the exponential map is $\exp(A) = e^{A}$.

Both for $GL(n, R)$ and for $SO(n)$, we see vividly that the exponential map is a local diffeomorphism.

For comic relief, we consider the examples $G = R^n$ and $G = T^n$, the n-torus. Since R^n and T^n are locally isomorphic, they have the same Lie algebra. We leave to the reader the task of verifying that the Lie algebra of R^n is R^n with the bracket product $[X, Y] \equiv 0$, and that the exponential is the identity map. The exponential map for T^n is the natural projection of R^n onto T^n. Note in passing that in T^n, some one-parameter groups are closed, while others are not. In R^n, however, all one-parameter subgroups are closed.

Section 6. Universal enveloping algebra

As we have already seen, the Lie algebra of a group G consists exactly of all left-invariant first-order differential operators on G which anihalate constants. We shall now study the *universal enveloping algebra* of G, which is just the (non-commutative) algebra of *all* left-invariant differential operators on G.

To be precise, let M be an n-manifold. A kth-order differential operator on M is a linear mapping $D: C^\infty(M) \to C^\infty(M)$ which can be written in terms of local co-ordinates $(x_1 \cdots x_n)$ for M defined in a neighborhood of $p \in M$, in the form $(Df)(x_1 \cdots x_n) = \Sigma_{|\alpha| \leq k}\, a_\alpha(x_1 \cdots x_n) \frac{\partial^\alpha f}{\partial x^\alpha}(x_1 \cdots x_n)$, for $f \in C^\infty(M)$, where a_α are fixed C^∞-functions defined in a neighborhood of p. A differential operator D on a Lie group G is said to be *left- invariant* if $D(\lambda_a f) = \lambda_a(Df)$ for any $a \in G$, $f \in C^\infty(G)$; where λ_a is the left-regular representation.

The left-invariant differential operators on G form an algebra (seldom commutative), which we denote by $\mathfrak{D}(G)$. For the third time, we note that every X belonging to the Lie algebra $\mathfrak{g}$ is a left-invariant differential operator, i.e., belongs to $\mathfrak{D}(G)$. It can be shown that $\mathfrak{g}$ generates the algebra $\mathfrak{D}(G)$. In fact we shall prove a far stronger result: Regard $\mathfrak{g}$ as a vector space, and let $T\mathfrak{g}$ denote the (non-commutative) tensor algebra of $\mathfrak{g}$, i.e., $T\mathfrak{g} = \Sigma_{k=0}^{\infty} \otimes^k \mathfrak{g}$. Equivalently, $T\mathfrak{g}$ is the algebra of real polynomials in the non-commuting variables $X_1 \cdots X_n$, where the X_i form a base for $\mathfrak{g}$. Let $\mathfrak{I}(G)$ denote the two-sided ideal in $T\mathfrak{g}$, generated by all expressions $X \otimes Y - Y \otimes X - [X, Y]$ where $X, Y \in \mathfrak{g}$. The quotient algebra $T\mathfrak{g}/\mathfrak{I}(G)$ is written $U(G)$ and is called the *universal enveloping algebra* of G. We shall exhibit a canonical isomorphism of $\mathfrak{D}(G)$ with $U(G)$.

THEOREM: $\mathfrak{g} \subseteq T\mathfrak{g} \to U(G)$ is an injection of $\mathfrak{g}$ into $U(G)$. If we identify $\mathfrak{g} \subseteq U(G)$ with $\mathfrak{g} \subseteq \mathfrak{D}(G)$, then the resulting correspondence extends uniquely to an isomorphism of $U(G)$ onto $\mathfrak{D}(G)$.

Proof: We can easily define an algebra homomorphism from $T\mathfrak{g}$ into $\mathfrak{D}(G)$, by mapping $X \otimes Y \otimes \cdots \otimes W$ into the differential operator

$$f \to X(Y(\cdots(Wf))\cdots), \text{ for } X, Y, \ldots, W \in \mathfrak{g} .$$

By definition of the bracket product, our homomorphism sends

$$X \otimes Y - Y \otimes X - [X, Y] \text{ to } 0 .$$

Hence we obtain a homomorphism j of algebras from the quotient U(G) into $\mathfrak{D}(G)$.

LEMMA: Let $X_1, \ldots, X_n$ be a base for $\mathfrak{g}$, and let $m = (m_1, \ldots, m_n)$ be a multi-index, $|m| = \Sigma_{k=1}^{n} |m_k|$. Define $X(m) \in U(G)$ to be the coefficient of $t_1^{m_1} \cdots t_n^{m_n}$ in the expression $\frac{1}{|m|!}(X_1 t_1 + \cdots + X_n t_n)^{|m|}$ (the t's are supposed to commute with each other and with the X's, but X_i and X_j do not commute). We set $X(0) \equiv 1$. Thus $X(m)$ is a "symmetrization" of $X_1^{m_1} \cdots X_n^{m_n}$. Then the elements $X(m)$ span the vector space U(G).

Proof: We must show that each monomial $Z = X_{k_1} X_{k_2} \cdots X_{k_\ell} \in U(G)$ is in the span of the $X(m)$. This is clear for $\ell \leq 1$, and we use induction on ℓ, the degree of Z. Suppose that appearing in Z are $m_1 X_1$'s, $m_2 X_2$'s, ..., $m_n X_n$'s. Set $m = (m_1, m_2, \ldots, m_n)$, so that $|m| = \ell$. It is not difficult to see that $Z = X(m) +$ lower-degree terms, in U(G). (This is because X(M) is a linear combination of all monomials obtained by rearranging the order of the X_{k_i} within Z. But any such monomial Z' differs from Z by terms of degree lower than that of Z. For example, consider the monomial $Z' = X_{k_2} X_{k_1} X_{k_3} X_{k_4} X_{j_5} \cdots X_{k_\ell}$. Then

$$Z - Z' = (X_{k_1} X_{k_2} - X_{k_2} X_{k_1}) X_{k_3} \cdots X_{k_\ell} = [X_{k_1}, X_{k_2}] \cdot X_{k_3} \cdot \cdots \cdot X_{k_\ell}$$

which has degree $\ell - 1$. The case of a general Z' is handled in the same way.) Since the lower degree terms are in the span of the $X(m)$ by inductive hypothesis, we have proved that Z is in the span of the $X(m)$. This completes the proof of the Lemma. QED

Now consider $j(X(m)) \in \mathfrak{D}(G)$. We show next that these elements are linearly independent. This will show that the $X(m)$ form a base for the vector space U(G), and that the map j is injective.

To prove the linear independence, we use canonical co-ordinates for the base $X_1 \cdots X_n$ of $\mathfrak{g}$. Let $f: G \to R$ be any real-analytic function.

Then in the spirit of formulas 1., 2., and 3. above, we can show that $F(t_1,\ldots,t_n) \equiv f(\exp(t_1X_1 + \cdots + t_nX_n)) = \Sigma_m t^m(jX(m)f)(x)$. Hence, in canonical co-ordinates $(t_1,\ldots,t_n)$, the elements $jX(m) \in \mathfrak{D}(G)$ correspond to the operators $\partial^{|m|}/\partial t_1^{m_1}\,\partial t_2^{m_2} \cdots \partial t_n^{m_n}$; and the latter are obviously linearly independent. So the $jX(m)$ are also linearly independent.

To finish the proof of the theorem, we have only to show that j is onto. Nothing could be simpler. First of all, a left-invariant differential operator D is uniquely determined by the functional $f \to (Df)(1)$. On the other hand, a monent's thought reveals that we can write

$$(Df)(0) = \sum_{|\alpha| \le k} a_\alpha X_{k_1^\alpha} \circ \cdots \circ Xf_{k_\ell^\alpha}(0),$$

(where a_α are fixed constants, and $X_{k_i^\alpha} \in \mathfrak{g}$) for *any* differential operator D. So if D is left-invariant,

$$Df = \sum_{|k| \le \alpha}^{k_i^\alpha} a_\alpha X_{k_1^\alpha} \cdots X_{k_\ell^\alpha} f .$$

Thus $D \in j(U(G))$, so that j is onto. QED

The center $Z(G)$, of the universal envoloping algebra $\mathfrak{D}$ consists of all those differential operators D which are *bi-invariant*, i.e., commute with both the left and the right regular representation. For suppose that D is bi-invariant. To show that $D \in Z(G)$, we need only check that D commutes with all $X \in \mathfrak{g}$, since $\mathfrak{g}$ generates the universal enveloping algebra. Any $X \in \mathfrak{g}$ may be written in the form $(Xf)(x) = \frac{d}{dt}(x \cdot \gamma(t))|_{t=0}$, where $\gamma(t)$ is the one-parameter group corresponding to X. Therefore

$$(X \circ D)f(x) = \frac{d}{dt}(Df)(x \cdot \gamma(t))\Big|_{t=0} = \frac{d}{dt}D(f(x \cdot \gamma(t))\Big|_{t=0}$$

$$= D\frac{d}{dt}f(x \cdot \gamma(t))\Big|_{t=0} = (D \circ X)f(x),$$

by the right-invariance, so X and D commute.

Conversely, suppose $D \in Z(G)$. With X and $\gamma(t)$ as above, we have

$$\frac{d}{dt} D \rho_{\gamma(t)} f \Big|_{t=0} = \frac{d}{dt} \rho_{\gamma(t)} Df \Big|_{t=0}$$

where ρ is the right regular representation; for this equation simply means $DXf = XDf$. Applying the last equation to the function $f = \rho_{\gamma(s)} g$, we obtain

$$\frac{d}{dt} D\rho_{\gamma(t)} g \Big|_{t=s} = \frac{d}{dt} \rho_{\gamma(t)} Dg \Big|_{t=s}$$

for any s. Since $D\rho_{\gamma(0)} = \rho_{\gamma(0)} D$, it follows that $D\rho_{\gamma(t)} g = \rho_{\gamma(t)} Dg$. This proves that D commutes with ρ_a if a is in the image of the exponential map, which contains a neighborhood N of 1 in G. Under our standing assumption that G is connected, we conclude that N generates G, which implies that D commutes with the right regular representation. So D is bi-invariant.

As a corollary to the above remark, we note that *the algebra of bi-invariant differential operators on* G *is commutative.*

A reasonable way to generalize non-trivial Fourier analysis to a Lie group G, would be to study the simultaneous spectral decomposition of all the bi-invariant differential operators. We shall not follow this approach strictly, but will rather study a single bi-invariant operator Δ, which will play the role of the Laplacian.

The base $\{X(m)\}$ for U(G) provides another useful isomorphism. In fact, let $\mathcal{S}(G)$ be the symmetric tensor algebra of $\mathfrak{g}$, i.e., the quotient algebra of $\mathcal{T}(G)$ with respect to the ideal generated by elements of the form $X \otimes Y - Y \otimes X$. Then $\{X(m)\}$ is also a base for $\mathcal{S}(G)$, as may be verified easily. So there is an induced isomorphism of vector spaces (not algebras) $\iota: \mathcal{S}(G) \to U(G)$ (onto). This isomorphism is actually independent of the base $X_1 \cdots X_n$ relative to which the constructions were made. We have thus the *Birkhoff-Witt Theorem*: $\mathcal{S}(G)$ is canonically isomorphic to $U(G)$.

The Birkhoff-Witt theorem allows us to formulate an interesting and important characterization of $Z(G)$, the bi-invariant operators, in terms of the *adjoint representation* of G, defined as follows. If $a \in G$, the inner automorphism $x \to a x a^{-1}$ of G, induces a linear mapping $\mathrm{Ad}(a)$ from the tangent space to G at 1, to itself. Thus $\mathrm{Ad}(a)\colon \mathfrak{g} \to \mathfrak{g}$. Since $\mathrm{Ad}(ab) = \mathrm{Ad}(a)\mathrm{Ad}(b)$ and $\mathrm{Ad}(1)$ is the identity mapping on $\mathfrak{g}$, we see that Ad is a representation of G on $\mathfrak{g}$; Ad is called the *adjoint representation.*

The adjoint representation allows us to make a "transition" between left and right multiplication in G. For example, $a \exp X = \exp(\mathrm{Ad}(a)X)a$.

Ad extends uniquely to a representation on $\mathcal{S}(G)$, the symmetric algebra—we just set $\mathrm{Ad}(a)(X \otimes Y \otimes \cdots \otimes W) = \mathrm{Ad}(a)X \otimes \mathrm{Ad}(a)Y \otimes \cdots \otimes \mathrm{Ad}(a)W$. Under the Birkhoff-Witt isomorphism i, the bi-invariant differential operators correspond precisely to those elements $Z \in \mathcal{S}(G)$ which are fixed by the transformations $\mathrm{Ad}(a)$. The easy proof, which uses the fact that $D \in \mathfrak{D}(G) = U(G)$ is uniquely determined by the functional $f \to Df(1)$, is left to the reader.

Section 7. Laplacian

We have now built up enough machinery to define a "Laplacian" on a general compact Lie group. Our Laplacian will mimic the behavior of the Laplacian $\Delta = \Sigma_{i=1}^{n} \partial^2/\partial x_i^2$ on R^n, in the following respects:

1. Δ is a second-order bi-invariant differential operator on R^n.
2. Δ has no zero-order term. In other words, Δ takes constant functions to be zero.
3. Δ is formally self-adjoint, This means that for any $f, g \in C_0^\infty(R^n)$,

$$\int_{R^n} f(x)\overline{\Delta g(x)}dx = \int_{R^n} \Delta f(x)\overline{g(x)}dx$$

which follows from integration by parts.

4. Δ is elliptic. In general, a second-order differential operator on R^n, given by $D = \Sigma_{i,j=1}^n a_{ij}(x) \dfrac{\partial^2}{\partial x_i\, \partial x_j}$ + lower order terms, with $a_{ij}(x) = \overline{a_{ji}(x)}$, is called *elliptic* if for any complex numbers $\xi_1, \ldots, \xi_n$, we have

$$\sum_{i,j=1}^{n} a_{ij}(x)\xi_i\bar{\xi}_j \geq c \sum_i |\xi_i|^2$$

with a fixed constant c, independent of x and ξ.

These concepts all make sense on a Lie group. Certainly the notion of a second-order bi-invariant differential operator presents no problems, and the condition $D(1) = 0$ is well-defined. We must be a little more careful with ellipticity and self-adjointness. A second-order differential operator D on a manifold M is called elliptic, if for any local co-ordinate on M, given by the diffeomorphism ϕ taking a neighborhood in M to a neighborhood U in R^n, the differential operator $\tilde{D}$ on $U \subseteq R^n$, defined by $\tilde{D}f \equiv (D(f \circ \phi)) \circ \phi^{-1}$, is elliptic. This definition agrees with our previous notion of ellipticity for open sets in R^n. For if ϕ is a diffeomorphism of the neighborhoods U and V of R^n, D is a second-order differential operator on V, and $\tilde{D}$ is the differential operator on U, defined by $\tilde{D}f \equiv (D(f \circ \phi)) \circ \phi^{-1}$; then D is elliptic if and only if $\tilde{D}$ is elliptic.

We have defined ellipticity of an operator on a C^∞-manifold, hence on any Lie group. Let us now turn to self-adjointness. A differential *operator* D on a Lie group G is *formally self-adjoint* if for any f, $g \in C_0^\infty(G)$,

$$\int_G f(x)Dg(x)dx = \int_G Df(x)\overline{g(x)}dx ,$$

where dx denotes left Haar measure on G. The subtlety is that if we used *right* Haar measure for dx, we would *not* end up with the same class of "formally self-adjoint operators." If G is compact, this problem disappears, since left and right Haar measure coincide.

THEOREM: Let G be a compact Lie group. There is a second-order differential operator D on G, such that

(a) D is bi-invariant;

(b) D is elliptic;

(c) D is formally self-adjoint;

(d) D maps constant functions to zero.

Moreover, D may be taken to be the form $\Sigma_{i,j=1}^{n} a_{ij} X_i X_j$, where $X_1 \cdots X_n$ form a base of the Lie algebra, and the constant matrix (a_{ij}) is strictly positive-definite.

Proof: Start with any operator $D_0 = \Sigma_{i,j=1}^{n} a_{ij}^0 X_i X_j$, where (a_{ij}^0) is a constant, strictly positive-definite matrix. D_0 belongs to the universal enveloping algebra, and corresponds under the Birkhoff-Witt isomorphism, with the symmetrization of the element $\Sigma_{i,j=1}^{n} a_{ij}^0 X_i \oplus X_j \in \mathcal{G}(G)$. But since (a_{ij}^0) is symmetric, $D^0 = \Sigma_{i,j=1}^{n} a_{ij}^0 X_i \oplus X_j$ is already symmetrized. So the Birkhoff-Witt isomorphism carries D_0 to $D^0 \in \mathcal{S}(G)$. If the transformation Ad(a) fixed D^0, for $a \in G$, then D_0 would be bi-invariant. In the general case, we make D^0 into an Ad(a)-invariant element by forming the "average," $D^1 = \int_G \mathrm{Ad}(x)(D^0)\,dx$. Here, dx is normalized so that $\int_G dx = 1$. This is the only place where we use compactness of G. By the usual computation, $\mathrm{Ad}(a) D^1 = D^1$. Each $\mathrm{Ad}(x)(D^0)$ is strictly positive-definite, since D^0 is; hence $D^1 = \Sigma_{i,j=1}^{n} a_{ij} X_i \oplus X_j \in \mathcal{S}(G)$ with (a_{ij}) strictly positive-definite. The Birkhoff-Witt isomorphism takes D^1 to the differential operator $D = \Sigma_{i,j=1}^{n} a_{ij} X_i X_j$, which must be bi-invariant, since $\mathrm{Ad}(a) D^1 = D^1$. D is elliptic, since in canonical co-ordinates it takes the form $\Sigma_{i,j=1}^{n} a_{ij} \dfrac{\partial^2}{\partial t_i\, \partial t_j}$. Clearly, D maps constant functions to zero. It remains only to show that D is formally self-adjoint. To do so, we shall need integration by parts on a Lie group. More pedantically, if $X \in \mathfrak{g}$ and $\rho, \psi \in C_0^\infty(G)$ then $\int_G (X\phi)\psi\,dx = -\int_G \phi(X\psi)\,dx$. The proof of the formula is quite simple. We set $\gamma(t) = \exp(tX)$, and differentiate the equation

$\int_G \phi(x)\psi(x)\,dx = \int_G \phi(x \cdot \gamma(t)\psi(x \cdot \gamma(t))dx$ in t . Setting $t = 0$, we obtain the desired formula.

Now for $\phi, \psi \in C^\infty(G)$ we have

$$\int_G D\phi(x)\overline{\psi(x)}\,dx = \sum_{i,j} a_{ij} \int_G X_i X_j \phi(x)\,\overline{\psi(x)}\,dx$$

$$= \sum_{i,j} a_{ij} \int_G \phi(x)\,\overline{X_j X_i \psi(x)}$$

$$= \int_G \phi(x)(\overline{\sum_{i,j} a_{ij} X_j X_i)\psi(x)}\,dx$$

$$= \int_G \psi(x)\,\overline{D\psi(x)}\,dx, \text{ dy}$$

the symmetry of (a_{ij}) and two applications of the integration-by-parts formula. Thus D is formally self-adjoint. QED

Only in rare cases is the operator D unique. On R^2, for instance, $(\partial^2/\partial x^2) + 3(\partial^2/\partial y^2)$ is just as good a Laplacian as $(\partial^2/\partial x^2) + (\partial^2/\partial y^2)$. It is not hard to show that D is unique up to constant multiple if and only if the Lie algebra $\mathfrak{g}$ has no proper "ideals." A Lie algebra (other than the trivial one-dimensional algebra) without proper ideals, is called a *simple* Lie algebra.

Henceforth, we shall work with a single differential operator Δ (on G), which satisfies the properties of D above. Δ will be called a *Laplacian* on the Lie group.

Problem: Show that Δ is the Laplace-Beltrami operator on G corresponding to a suitably defined bi-invariant Riemannian metric on G. (For the properties of the Laplace-Beltrami operator see e.g., Helgason [2, Chapter X].)

CHAPTER II

LITTLEWOOD-PALEY THEORY FOR A COMPACT LIE GROUP

In this chapter we present the analogue of the Littlewood-Paley theory for any compact Lie group. This development is independent of the more general one given in Chapters III and IV, but has distinct advantages of its own. It is based, in effect, is the following special circumstances. The functions $u(x, t) = (P^t f)(x)$, where P^t is the Poisson semi-group are solutions of the equation $\Delta u(x, t) = 0$; various special properties of the Laplacian Δ come into play (for more details on this point see Section 3 below).

The special approach followed here can also be used in other contexts, such as the case of non-compact semi-simple groups G treated in Chapter V.

The present chapter is organized as follows. Section 1 constructs the heat-diffusion semi-group on G, representing the solutions of the heat-equation $\partial u/\partial t = \Delta u$. Among the general semi-groups that we consider later in Chapters III and IV, this gives probably the simplest and most easily constructed example. In Section 2 we pass from the heat semi-group, to the Poisson semi-group, (which represents the solutions of $(\partial^2 u/\partial t^2) + \Delta u = 0$), by a familiar process fo "subordination." In terms of this latter semi-group we define the g-function and study its properties in Section 3. Various applications are pointed out in Section 4.

Section 1. The heat-diffusion semi-group

Our construction of the Poisson integral will be made in terms of a simpler object, namely the semigroup $\{T^t\}$ of operators defined by $T^t = e^{t\Delta}$, in a sense which we shall make precise. (Δ denotes the Laplacian.) More rigorously, we shall show

THEOREM 1: *There exist operators* T^t, $0 \leq t < +\infty$, *such that*

i) *each* T^t *is a bound operator of norm* 1 *on* $L_p(G)$, $1 \leq p \leq +\infty$ *and also on* $C(G)$;

ii) $\{T^t\}$ *form a semigroup, i.e.,* $T^{t_1} T^{t_2} = T^{t_1+t_2}$, *and* T^0 *is the identity operator;*

iii) $\{T^t\}$ *is strongly continuous, i.e., the mapping* $t \to T^t f$ *is continuous from* $[0, \infty)$ *into* $L_p(G)$, *for each fixed* $f \in L_p(G)$ $(p \neq \infty)$. *In the case* $p = +\infty$ *we have instead that* $t \to T^t f$ *mapping* $[0, \infty)$ *into* $C(G)$ *is continuous, for each* $f \in C(G)$;

iv) *each* T^t *is a self-adjoint operator on* $L_2(G)$;

v) T^t *is positive, i.e.,* $f \geq 0$ *implies* $T^t f \geq 0$;

vi) $T^t 1 = 1$;

vii) *the Laplacian* Δ *is the infinitessimal generator of the semigroup* $\{T^t\}$, *i.e.,* $\lim_{t\to 0}(T^t - I)f/t = \Delta f$, *whenever* f *belongs to a suitable dense subset of* $C(G)$;

viii) T^t *may be written in the form* $T^t f = K_t * f$, $t > 0$, *where* $K_t(x)\colon G \times (0, \infty) \to R$ *is a* C^∞ *function;*

ix) *if for* $f \in L_1(G)$ *we set* $u(x, t) = (T^t f)(x)$, *then* $u(x, t) \in C^\infty(G \times (0, \infty))$ *satisfies the heat equation* $\frac{\partial}{\partial t} u(x,t) = \Delta u(x,t)$ *and the boundary condition* $u(\cdot\,, t) \to f$ *in* $L_1(G)$ *as* $t \to 0+$.

Condition ix) is the raison-d'etre of the semigroup $\{T^t\}$. As soon as we have proved Theorem 1, we shall be able to write down an expression for the Poisson integral in terms of $\{T^t\}$.

Much machinery could be brought to bear in the proof of Theorem 1. There is an extensive theory of semigroups from which it is possible to prove the existence (and uniqueness) of $\{T^t\}$ satisfying conditions (i)-(viii) above. The smoothness conditions (viii) and (ix) really express the fact that the heat equation is hypoelliptic.

We shall not have to use any of this machinery because we have concrete information, namely an explicit eigenfunction expansion of Δ (see below), which will enable us to give an elementary proof of Theorem 1. But using the theory of semigroups and of partial differential equations, we can generalize Theorem 1. from a Lie group to a Riemannian manifold M. In this case, Δ is the Laplace-Beltrami operator on M, and again there is a semigroup $\{T^t\}$ as above such that $T^t f(x) = u(x,t)$ solves the heat equation.

For the general theory of semigroups, see Feller [10], Yosida [40], and Hille-Phillips [11]. Semigroups of operators arising from Lie groups are treated in the papers *Semigroups of Measures on Lie Groups* by G. Hunt (*T.A.M-S.*, 1956) [12], and *Analytic Vectors*, by E. Nelson (*Ann. Math.*, 1959), [14].

Proof of Theorem 1: We shall first produce an eigenfunction expansion of Δ, using the Peter-Weyl theorem. For purposes of this proof we use the (slightly non-standard) terminology of an *eigenfunction* ϕ of Δ with *eigenvalue* λ, which means $\Delta\phi = -\lambda\phi$ (instead of the usual $\lambda\phi$). With this convention, all eigenvalues of Δ are non-negative, because, $(\Delta f, f) \le 0$ for any $f \in C^\infty(G)$, $(\cdot, \cdot)$ denoting the usual inner product on $L_2(G)$. In fact,

$$(\Delta f, f) = \int_G \Big(\sum_{ij} a_{ij} X_i X_j f\Big)(x)\overline{f(x)}\,dx$$

$$= \sum_{ij} a_{ij} \int_G X_i X_j f(x)\overline{f}(x)\,dx$$

$$= -\sum_{ij} a_{ij} \int_G X_j f \cdot \overline{X_i f}\,dx = -\int_G \sum_{ij} a_{ij} \overline{X_i f} X_j f\,dx \le 0,$$

with equality if and only if $X_i f = 0$, i.e., $f =$ constant. (We have used the strict positivity of the matrix (a_{ij})).

Now recall the Peter-Weyl decomposition $L_2(G) = \Sigma_{\alpha \epsilon \Lambda} \oplus H_\alpha$, where Λ denotes the set of equivalence classes of finite-dimensional irreducible representations of G, and H_α denotes the vector space spanned by the (degree $\alpha)^2$ entries ϕ_{ij} of a representation $R_\alpha = \{\phi_{ij}(x)\}$ of class α. Each H_α is contained in $C^\infty(G)$; so it makes sense to take the Laplacian of any function in $\mathfrak{E}$, the algebraic direct sum of the H_α. We shall prove that each $\phi \epsilon H_\alpha$ is an eigenvector of Δ, and that all $\phi \epsilon H_\alpha$ have the same eigenvalue λ_α.

In fact, let $\psi_{ij} = \Delta\phi_{ij}$, where $\{\phi_{ij}(x)\}$ is a representation of class α. Then since $\phi_{ij}(xa) = \Sigma_k \phi_{ik}(x)\phi_{kj}(a)$, we have $\psi_{ij}(xa) = \Sigma_k \psi_{ik}(x)\phi_{kj}(a)$ by the right-invariance of Δ. Setting $x = 1$, we obtain $\psi_{ij}(y) = \Sigma_k \psi_{ik}(1)\phi_{kj} \cdot (y)$. Similarly $\phi_{ij}(ax) = \Sigma_k \phi_{ik}(a)\phi_{kj}(x)$ so $\psi_{ij}(y) = \Sigma_k \phi_{ik}(y)\psi_{kj}(1)$, by the left-invariance of Δ. Thus, the matrix $\{\psi_{ij}(1)\}$ commutes with the irreducible representation $\{\phi_{ij}(x)\}$, so that by Schur's lemma, $\{\psi_{ij}(1)\} = -\lambda_\alpha I$, where I denotes the identity matrix. We conclude that $\psi_{ij}(y) = \Sigma_k \phi_{ik}(y)\psi_{kj}(1) = -\lambda_\alpha\phi_{ij}(y)$. which shows that $\Delta\phi = -\lambda_\alpha\phi$ for $\phi \epsilon H_\alpha$. Recall that λ_α must be non-negative.

The decomposition $L_2(G) = \Sigma_{\alpha \epsilon \Lambda} \oplus H_\alpha$ is therefore an eigenfunction decomposition with respect to Δ. For convenience, we pick orthonormal bases of all the H_α, and list all the base elements $1, \phi_1(x), \phi_2(x), \ldots$. Each ϕ_k is an eigenfunction of Δ, with eigenvalue μ_k (note that $\mu_0 = 0$, and $\mu_k > 0$ for $k \geq 1$).

The space $\mathfrak{E}$ consists exactly of all finite linear combinations of the ϕ_i. If $f = \Sigma\, a_i\phi_i \,\epsilon\, \mathfrak{E}$ (a_i finitely non-zero) then of course $\Delta f = \Sigma - \mu_i a_i \phi_i$. This suggests a definition for $T^t = e^{t\Delta}$: for $f = \Sigma\, a_i\phi_i \;\epsilon\; \mathfrak{E}$, we set $T^t f = \Sigma\, e^{-\mu_i t} a_i\phi_i$. Obviously $\|T^t f\|_2^2 = \Sigma_i |e^{-\mu_i t} a_i|^2 \leq \Sigma_i |a_i|^2 = \|f\|_2^2$ for $f \epsilon \mathfrak{E}$, since $\mu_i \geq 0$ and $t \geq 0$. Thus T^t extends from $\mathfrak{E}$, to a bounded linear operator on $L_2(G)$ of norm 1—the extension is also denoted by T^t. The reader may check that $\{T^t\}$ is a strongly continuous semigroup of self-adjoint operators on L_2, $T^t 1 = 1$; and if $f \epsilon \mathfrak{E}$, the function

$u(x, t) = (T^t f)(x)$ belongs to $C^\infty(G \times (0, t))$ and satisfies the heat equation. Hence, the L_2-theory of the operators T^t is well in hand.

Next, we show that T^t is a positive operator. From this fact the L_p-properties of T^t will be easy.

Define the *resolvent* $R(\lambda, \Delta)$ to be the operator $(\lambda I - \Delta)^{-1}$, for $\lambda > 0$. The analysis of $R(\lambda, \Delta)$ trivializes here, for $f = \Sigma\, a_i \phi_i(x) \in \mathfrak{S}$,

$$R(\lambda, \Delta) f = \sum \frac{1}{(\lambda + \mu_i)}\, a_i \phi_i(x) \;;$$

which shows when $\lambda > 0$, $R(\lambda, \Delta)$ is a bounded operator on $L_2(G)$. This is typical of the usefulness of $\mathfrak{S}$ in avoiding all technical difficulties.

A standard fact from semigroup theory is that each T^t is positive if and only if $R(\lambda, \Delta)$ is positive for $\lambda > 0$. On the one hand,

$$\text{(A)} \qquad T^t f = \lim_{n \to \infty} \left(\frac{n}{t} R\left(\frac{n}{t}, \Delta\right)\right)^n f \qquad \text{for} \qquad f \in \mathfrak{S},$$

since we can write $f = \Sigma\, a_i \phi_i(x)$ (finite sum) and then equation (A) reduces to

$$\sum e^{-\mu_i t} a_i \phi_i(x) = \lim_{n \to \infty} \sum \left(\frac{n/t}{(n/t) + \mu_i} \right)^n a_i \phi_i(x) \;,$$

So if $R(\lambda, \Delta)$ is positive, then $(\frac{n}{t} R(\frac{n}{t}, \Delta))^n$ is positive, which implies that T^t is positive. Similarly, that $T^t \geq 0$ implies $R(\lambda, \Delta) \geq 0$, follows from the identity

$$\text{(B)} \qquad R(\lambda, \Delta) f = \int_0^\infty e^{-\lambda t}\, T^t f\, dt \qquad (f \in \mathfrak{S}),$$

which in turn comes from the same kind of routine computation as (A).

We can now show that $R(\lambda, \Delta)$ is positive on $\mathfrak{S}$. Let $f \in \mathfrak{S}$, and suppose that $f \geq 0$ and $R(\lambda, \Delta) f = g \in \mathfrak{S}$. Then $\lambda g - \Delta g = f \geq 0$. We must show $g \geq 0$. If this did not hold, then at the point $x_0 \in G$ at which g is a minimum (recall that G is compact) we have $g(x_0) < 0$. On the other hand,

$\Delta g(x_0) \geq 0$ since g takes its minimum at x_0. Hence $\lambda g(x_0) - \Delta g(x_0) < 0$, contradicting $f \geq 0$. This completes the proof that $R(\lambda, \Delta)$ and therefore T^t is positive. We have used tacitly the fact that $R(\lambda, \Delta)f$ is real if f is real, which we leave as an exercise to the reader.

Since the positive operator $T^t: \mathcal{E} \to \mathcal{E}$ maps 1 into 1, it follows that T^t extends to a positive $T^t: C(G) \to C(G)$ of norm 1. For fixed $t > 0$ and $x_0 \epsilon G$, the positive linear functional $f \epsilon C(G) \to (T^t f)(x_0)$ is of the form

$$(T^t f)(x_0) = \int_G f(x) \, d\mu^t_{x_0}(x)$$

where $d\mu^t_{x_0}$ is a positive measure with total mass 1, by the Riesz representation theorem. On the other hand, the operator T^t is bi-invariant, since for $f \epsilon \mathcal{E}$, $T^t f = \Sigma \ t^n \Delta^n f / n!$. Therefore $\mu^t_{x_0}(E) = \mu^t_1(x_0^{-1} E)$, which means that $T^t f(x_0) = (f * \mu^t_1)(x_0)$ for every $f \epsilon C(G)$. Since

$$\| T^t f \|_p = \| \int_G f(y^{-1}x) \, d\mu_1(y) \|_p \leq \int_G \| f(y^{-1} \cdot) \|_p \, d\mu_1(y)$$

$$= \| f \|_p \int_G d\mu_1(y) = \| f \|_p \ ,$$

we have verified property i) in the statement of the theorem.

Property iii) follows from property i) and the density of $\mathcal{E}$ in L_p. It remains to check properties viii) and ix). Now ix) is clear in the case $f \epsilon \mathcal{E}$, from which we deduce by a routine limiting argument that $u(x, t) = (T^t f)(x)$ satisfies the heat equation, for any $f \epsilon L_1$, once we have proved viii).

To prove viii) we require a simple form of the Sobolev lemma; which we state as an a priori inequality: Let $f \epsilon C_0^\infty(R^n)$. Then

$$\sup_{x \in R^n} |f(x)| \leq A \sum_{0 \leq |\alpha| \leq N} \left\| \frac{\partial}{\partial x^\alpha} f \right\|_2 ,$$

where N is any integer > n/2, and A depends only on n.

The lemma is most easily proved by means of the Plancherel theorem. In fact, $\|f\|_\infty \leq \|\hat{f}\|_1 \leq \|(1+|y|)^{-N}\|_2 \, \|(1+|y|)^N \hat{f}(y)\|_2$

$$\leq K \sum_{|\alpha| \leq N} \| \, |y|^\alpha \hat{f}(y)\|_2 = K \sum_{|\alpha| \leq N} \left\| \frac{\partial}{\partial x^\alpha} f \right\|_2 .$$

Next, we shall extend our Lemma, and transfer it to the setting of the compact Lie group G:

LEMMA. *Let* $f \in C^\infty(G)$. *Then*

(a) $$\|f\|_\infty \leq A \sum_{\ell=0}^{N} \|\Delta^\ell f\|_2$$

where A *depends only on* G, *and* N *is any integer* > n/2.

To prove this, first observe that

(b) $$\|X_k f\|_2 \leq C \|\Delta f\|_2^{1/2} \, \|f\|_2^{1/2} ,$$

where X_k belongs to our basis for the Lie algebra $\mathfrak{g}$. For $\|\Delta f\|_2 \|f\|_2 \geq -(\Delta f, f) = \Sigma_{ij} a_{ij}(X_j f, X_i f) \geq C\|X_k f\|_2^2$ by the strict positivity of (a_{ij}). Combining (b) with the inequality $2ab \leq a^2 + b^2$, we obtain $\|X_k f\|_2 \leq C(\|\Delta f\|_2 + \|f\|_2)$. Repeated application of this inequality yields

$$\|P(X_1 \cdots X_n) f\|_2 \leq C'_P \sum_{\ell=0}^{N} \|\Delta^\ell f\|_2$$

for P(X) in the universal enveloping algebra of G, of degree ≤ N. To complete the proof of (a), we have only to show that $\|f\|_\infty \leq \Sigma_{p \in A} \|P(X_1 \cdots X_n) f\|_2$ for some finite A of P(X) of degree at most N. If f has small enough support, this follows from our Sobolev lemma and an application of canonical co-ordinates—the general case then follows from a partition of unity argument. Thus, inequality (a) holds.

Let us apply the above estimates to the study of the operator T^t and ist eigenvectors ϕ_k. Since $\Delta\phi_i = -\mu_i\phi_i$ and $\|\phi_i\|_2 = 1$, inequality (a) shows that $\sup_{x \in G} |\phi_i(x)| \leq C(1 + |\mu_i|)^N$ for any $N > \frac{1}{2} \dim G$. Consider any $f = \Sigma a_k \phi_k(x) \in \mathfrak{E}$, and let $u(x,t) = T^t f(x) = \Sigma e^{-\mu_k t} a_k \phi_k(x)$. Then $\Delta^\ell u(x,t) = \Sigma(-\mu_k)^\ell e^{-\mu_k t} a_k \phi_k(x)$, so that

$$\|\Delta^\ell u(\cdot,t)\|_2^2 = \Sigma \mu_k^{2\ell} e^{-2\mu_k t} |a_k|^2 \leq Ct^{-2\ell} \Sigma |a_k|^2 = Ct^{-2\ell}\|f\|_2^2 .$$

(We make use of the elementary inequality $\mu^{2\ell} e^{-2\mu t} \leq Ct^{-2\ell}$, valid for $\mu, t \geq 0$). We have proved an a priori estimate for $\|\Delta^\ell u(\cdot,t)\|_2$. Since $(\partial/\partial t)u(x,t) = \Delta u(x,t)$ for $f \in \mathfrak{E}$, we also obtain the a priori inequality

$$\left\| \frac{\partial^k}{\partial t^k} \Delta^\ell u(x,t) \right\|_{L_2(G)} \leq Ct^{-M}\|f\|_2$$

where C and M depend on k and ℓ alone. By the Sobolev lemma and familiar limiting arguments, $u(x,t) = (T^t f)(x)$ belongs to $C^\infty(G \times (0, \infty))$ for every $f \in L_2(G)$, and the map $f \to u(x, t)$ is a continuous operator from $L_2(G)$ to $C^\infty(G \times (0, \infty))$.

Property viii) above is now easy to prove. For, as a very special case of what we just showed, we have $T^t: L_2(G) \to L_\infty(G)$ is a bounded operator. The usual duality argument shows that T^t is also bounded as a mapping from $L_1(G)$ into $L_2(G)$. But then $T^t = T^{t/2}T^{t/2}$, the composition of continuous operators from $L_1 \to L_2$ and from $L_2 \to C^\infty$. Thus, for any $f \in L_1(G)$, $T^t f(x) \in C^\infty(G \times (0, \infty))$, which completes the proof of Theorem 1. QED

We can even give an explicit representation for the operator T^t, in terms of the docomposition $L_2(G) = \Sigma_{\alpha \in \Lambda} \oplus H_\alpha$. As usual, let us select a unitary representation $\{\phi_{ij}^\alpha(x)\}$ of class α, for each $\alpha \in \Lambda$, and use as our orthonormal base $\{\phi_i\}$, the family of vectors $\{d_\alpha^{1/2} \phi_{ij}^\alpha\}_{\alpha, i, j}$, where the factor $d_\alpha^{1/2} = (\text{degrees } \alpha)^{1/2}$ is put in to normalize vectors. The eigenvalue of Δ corresponding to $d_\alpha^{1/2}\, \phi_{ij}^\alpha$ is λ_α in our previous notation. Hence $T^t(d_\alpha^{1/2}\, \phi_{ij}^\alpha) = e^{-\lambda_\alpha t} d_\alpha^{1/2}\, \phi_{ij}^\alpha$, so we can write formally,

(c) $$T^t f(x) = \int_G (\Sigma_\alpha e^{-\lambda_\alpha t} d_\alpha \chi_\alpha(xy^{-1})) f(y)dy \ ,$$

since

$$\Sigma_{i,j,\alpha} e^{-\lambda_\alpha t} d_\alpha^{1/2}\, \phi_{ij}^\alpha(x) \cdot d_\alpha^{1/2} \phi_{ij}(y) = \Sigma_\alpha e^{-\lambda_\alpha t} d_\alpha\ \Sigma_{i,j}\, \phi_{ij}^\alpha(x)\phi_{ji}^\alpha(y^{-1})$$

$= \Sigma_\alpha\, e^{-\lambda_\alpha t} d_\alpha \chi_\alpha(xy^{-1})$, which follows because $\{\phi_{ij}^\alpha(x)\}$ is unitary, with character χ_α. We shall verify that this identity holds, not just formally, but literally, by showing that the series

$$\Sigma_\alpha\, e^{-\lambda_\alpha t} d_\alpha \chi_\alpha(x)$$

converges in the strongest possible sense. Namely any order partial derivative, with respect to x and t, of the series $\Sigma_\alpha\, e^{-\lambda_\alpha t} d_\alpha\, \chi_\alpha(x)$ converges absolutely absolutely and uniformly for $x \in G$, $t \geq \delta > 0$. In fact, since $f \to \Delta^\ell\, T^t f(1)$ is a bounded linear functional on $L_2(G)$ for each ℓ and since (c) holds formally, we have

$$\sum_{\alpha \in \Lambda} \lambda_\alpha^N\, e^{-2\lambda_\alpha t}\, d_\alpha^2 < +\infty$$

for each $N > 0$. Since $t > 0$ is arbitrary, $d_\alpha \geq 1$, and $\chi_\alpha(\cdot)$ has C^k-norm at most some fixed power of λ_α (again by our form of the Sobolev lemma), we conclude that

$$\sup_{s \geq \delta > 0} \sum_{\alpha \in \Lambda} e^{-\lambda_\alpha s}\, d_\alpha \|\chi_\alpha\|_{C^k(G)} < \infty$$

for each k, which implies the desired conclusion.

Section 2. The Poisson semi-group; the main theorem

As indicated above, we are not mainly concerned with $\{T^t\}$, but rather with the *Poisson* semigroup $\{P^t\}$, which is defined roughly as saying that $P^t f(x) = u(x, t)$ is the solution of Laplace's equation $((\partial^2/\partial t^2) + \Delta)u = 0$ with the boundary condition $u(x, 0) = f(x)$. It is easy to find such a u for $f \in \mathfrak{E}$. In fact, if $f = \Sigma_k a_k \phi_k \in \mathfrak{E}$, we can set

$$P^t f = \sum_k e^{-(\mu_k)^{1/2} t} \cdot a_k \phi_k \qquad (t \geq 0) ,$$

where the notation is as in the proof of Theorem 1. Thus, formally,

$$P^t = e^{-t(-\Delta)^{1/2}} .$$

THEOREM 1′. *The operator* P^t *just defined on* $\mathfrak{E}$ *extends to a bounded operator on* L_p $(1 \leq p \leq +\infty)$ *and on* C(G). *Moreover*

i) $\|P^t f\|_p \leq \|f\|_p$, *for all* f $(t \geq 0)$;

ii) P^t *is a self-adjoint operator on* $L_2(G)$;

iii) $f \geq 0$ *implies* $P^t f \geq 0$;

iv) $\lim_{t \to 0} \dfrac{P^t f - f}{t} = (-\Delta)^{1/2} f$ *for* $f \in \mathfrak{E}$. $(-\Delta)^{1/2}$ *is defined by* $(-\Delta)^{1/2} \phi_k = \mu_k^{1/2} \phi_k$;

v) $u(x, t) \equiv P^t f(x)$ *belongs to* $C^\infty(G \times (0, \infty))$ *and satisfies the equation* $((\partial^2/\partial t^2) + \Delta)u = 0$ *and the boundary condition* $u(x,t) \to f(x)$ *in the* $L_1(G)$ *norm if* $f \in L_1(G)$.

Proof: We shall prove the properties of P^t from analogous properties of T^t. by means of the heuristic *principle of subordination.* From the well-known identity

$$e^{-\beta} = \frac{1}{\sqrt{\pi}} \int_0^\infty \frac{e^{-u}}{\sqrt{u}} e^{-\beta^2/4u} \, du$$

valid for $\beta > 0$,

$$(*)\qquad P^t f = \frac{1}{\sqrt{\pi}} \int_0^\infty \frac{e^{-u}}{\sqrt{u}} T^{t^2/4u} f\, du \qquad (t > 0)\ .$$

Hence

$$\|P^t f\|_p = \left\| \frac{1}{\sqrt{\pi}} \int_0^\infty \frac{e^{-u}}{\sqrt{u}} T^{t^2/4u} f\, du \right\|_p \leq \frac{1}{\sqrt{\pi}} \int_0^\infty \frac{e^{-u}}{\sqrt{u}} \|T^{t^2/4u} f\|_p\, du$$

$$\leq \|f\|_p \left(\frac{1}{\sqrt{\pi}} \int_0^\infty \frac{e^{-u}}{\sqrt{u}}\, du \right) = \|f\|_p\ ,$$

which shows that P^t extends to a bounded operator on $L_p(G)$ and that (*) holds for all $f \in L_p(G)$. Properties i, ii, iii, and v follow from the corresponding facts about T^t and (*). QED

The function $P^t f(x)$ is called the *Poisson integral* of f, just as in the case G = the circle group.

Using the Poisson integral, we can formulate and prove a generalization of the classical Littlewood-Paley inequality for the g-function.

Recall that for f, say, a real-valued function in $C^\infty(G)$, $\Delta f = \Sigma_{ij}\, a_{ij} X_i X_j f$. Define $|\nabla f|^2(x)$ to be $\Sigma_{ij}\, a_{ij}(X_i f)(X_j f)$. Similarly, for $f \in C^\infty(G \times (0, \infty))$ real-valued, set $\boldsymbol{\Delta} f(x,t) = (\partial^2 f/\partial t^2) + \Delta_x f$ and $|\boldsymbol{\nabla} f|^2(x,t) = ((\partial/\partial t) f)^2 + |\nabla_x f|^2$.

Now let $f \in L_p(G)$, and let $u(x, t)$ be its Poisson integral. Then we define the *Littlewood-Paley g-function* for f to be

$$g(f)(x) \equiv \left(\int_0^\infty t\, |\boldsymbol{\nabla} u(x,t)|^2\, dt \right)^{1/2}\ .$$

THEOREM 2: *If* $f \in L_p(G)$ $(1 < p < +\infty)$, *then* $g(f)$ *is also in* $L_p(G)$, *and have the inequality* $\|g(f)\|_p \leq A_p \|f\|_p$ *where* A_p *depends only on* p. *Conversely, if* $\int_G f(x)\,dx = 0$, *then* $\|f\|_p \leq B_p \|g(f)\|_p$.

Proof: Note first that $\int_G f(x)dx = 0$ or something similar is needed if $\|f\|_p \leq B_p\|g(f)\|_p$ is to hold, for the g-function of every constant is zero.

Our proof develops a recent approach for the classical case, and requires a series of Lemmas, the first of which generalizes the Hardy-Littlewood maximal theorem.

LEMMA 1: *For any (reasonable) function* f, *define the maximal function* f^* *as* $f^*(x) = \sup_{t>0}|P^t f(x)|$. *Then if* $1 < p \leq +\infty$, $\|f\|_p \leq A_p\|f\|_p$ *for every* $f \in L_p(G)$, *with* A_p *independent of* f.

Proof of Lemma 1: The sledge-hammer with which we crack this problem is the Hopf-Dunford-Schwartz ergodic theorem, one of the most powerful results in abstract analysis. The form in which we shall use the theorem is this: Let $\{T^t\}$ be a (measurable) semigroup of operators on $L_p(\mathfrak{M}, d\mu)$. (Here $(\mathfrak{M}, d\mu)$ denotes an arbitrary measure space.) Suppose that $\|T^t f\|_p \leq \|f\|_p$ for all $f \in L_p(\mathfrak{M}, d\mu)$, and for each $p \in [1, \infty]$. Then the "maximal function" Mf, defined as

$$Mf(x) = \sup_{s>0}\left(\frac{1}{s}\int_0^s |T^t f(x)|\,dt\right)$$

satisfies the inequalities

(a) $\|Mf\|_p \leq A_p\|f\|_p$ for each p with $1 < p \leq +\infty$;

(b) $\mu(\{x \in \mathfrak{M} |\ Mf(x) > a\}) \leq \frac{A}{a}\|f\|_1$ for each $a > 0$ and $f \in L_1$, where A is independent of f and a.

The proof of this theorem is in Dunford-Schwartz [9] *Linear Operators*, Chapter VIII.

Although (b) is the basic inequality of the theorem, we shall only use (a) here. For $(\mathfrak{M}, d\mu)$ we take G with its Haar measure; for T^t we take the heat semigroup of Theorem 1. Inequality (a) gives us control of the averages of $T^t f$. This is exactly what we need, since (*) in the proof of Theorem 1′, shows that P^t is nothing but a weighted average of the T^t. In fact, changing variables in (*), we obtain

$$(**)\qquad P^t f = t^{-2}\int_0^\infty \phi\left(\frac{y}{t^2}\right) T^y f\, dy$$

where

$$\phi(y) \equiv \frac{1}{2\sqrt{\pi}} e^{-1/4y} y^{-3/2} .$$

The reader may verify that $\phi(y)$ and $y\phi'(y)$ belong to $L_1(0,\infty)$.

Integration by parts in (**) shows that

$$|P^t f(x)| = \left|-t^{-2}\int_0^\infty \left(\int_0^y T^t f(x)dt\right)\frac{d\phi}{dy}(y/t^2)dy\right|$$

$$= \left|-t^{-2}\int_0^\infty \left(\frac{1}{y}\int_0^y T^t f(x)dt\right)\left(y\frac{d}{dy}\phi\left(\frac{y}{t^2}\right)\right)dy\right|$$

$$\leq \sup_{y>0}\left|\frac{1}{y}\int_0^y T^t f(x)dt\right| \cdot \left(t^{-2}\int_0^\infty \left|y\frac{d\phi}{dy}\left(\frac{y}{t^2}\right)\right|dy\right)$$

$$\leq A \sup_{y>0}\left|\frac{1}{y}\int_0^y T^t f(x)dt\right| ,$$

where $A = \|y\phi'(y)\|_1$. Hence $P^t f(x) \leq AMf(x)$ for each $t > 0$, so that $f^*(x) \leq AMf(x)$. The lemma now follows from inequality a). QED

LEMMA 2: *Let* $u(x,t)$ *be a harmonic function on* $G \times (0,\infty)$, *i.e.,* $\Delta u = 0$; *and suppose that* $u > 0$. *Then for any* $p \geq 1$, $\Delta u^p = p(p-1)u^{p-2}|\nabla u(x,t)|^2$.

Proof: Since X_j corresponds to $\partial/\partial t_j$ in canonical co-ordinates, we have $X_j u^p = pu^{p-1}X_j u$. So $X_i X_j u^p = pu^{p-1}X_i X_j u + p(p-1)u^{p-2}(X_i u)(X_j u)$. Similarly $(\partial^2/\partial t^2)u^p = pu^{p-1}(\partial^2/\partial t^2)u + p(p-1)u^{p-2}(\partial u/\partial t)^2$. Combining these identities, we obtain

$$\Delta u^p = \frac{\partial^2}{\partial t^2}u^p + \sum_{ij} a_{ij}X_i X_j u^p = pu^{p-1}\left(\frac{\partial^2 u}{\partial t^2} + \sum_{ij} a_{ij}X_i X_j u\right)$$

$$+ p(p-1)u^{p-2}\left(\left|\frac{\partial u}{\partial t}\right|^2 + \sum_{ij} a_{ij}(X_i u)(X_j u)|\right)$$

$$= pu^{p-1}\Delta u + p(p-1)u^{p-2}|\nabla u|^2 = p(p-1)u^{p-2}|\nabla u|^2 . \qquad \text{QED}$$

LEMMA 3: *Let* $F(x,t) \in C^\infty(G \times [0,\infty])$ *and assume that* $t|\frac{\partial}{\partial t}F| \to 0$ *as* $t \to 0$ *or as* $t \to +\infty$. *Then*

$$\int_0^\infty \int_G \Delta F(x,t)\,dx\,dt = \int_G F(x,0)\,dx - \int_G F(x,+\infty)\,dx\ .$$

Proof: For any $f \in C^\infty(G)$,

$$\int_G X_j f(x)\,dx = \int_G 1 \cdot X_j f(x)\,dx = -\int_G (X_j 1) \cdot f(x)\,dx = 0\ .$$

Hence

$$\int_0^\infty t \int_G \Delta F(x,t)\,dxdt = \int_0^\infty t \int_G \frac{\partial^2 F}{\partial t^2}(x,t)\,dxdt\ ,$$

so Lemma 3 is obtained by integrating the identity

$$(***) \qquad \int_0^\infty t\,\frac{\partial^2}{\partial t^2}F(x,t)\,dt = F(x,0) - F(x,\infty)$$

over the group G. To prove (***), we integrate by parts, finding

$$\int_0^\infty t\,\frac{\partial^2}{\partial t^2}F(x,t)\,dt = t\frac{\partial F}{\partial t}\Big|_0^\infty - \int_0^\infty \frac{\partial F}{\partial t}(x,t)\,dt = F(x,0) - F(x,+\infty)\ .$$

QED

Section 3. Proof of Theorem 2

Now we can proceed to the proof of Theorem 2.

Part I: We will prove that $\|g(f)\|_p \le A_p\|f\|_p$, for $1 < p \le 2$. To do so, we need only consider $f \in C^\infty(G)$ which are strictly positive. $f \ge \varepsilon > 0$. Since $f \in C^\infty$ (we could also get away with assuming $f \in \mathfrak{S}$) the Poisson integral $u(x,t) \in C^\infty(G \times [0,\infty])$. In addition, $u(x,t) \ge \varepsilon > 0$, so that $F(x,t) = (u(x,t))^p$ also belongs to $C^\infty(G \times [0,\infty])$. The verification that

$t\partial F/\partial t\,(x,t) \to 0$ as $t \to 0$ or as $t \to +\infty$ is left to the reader. Lemma 2 shows that $\Delta F \geq 0$.

Now $(g(f)(x))^2 = \int_0^\infty t|\nabla u|^2 dt = \frac{1}{p(p-1)} \int_0^\infty t\, u(x,t)^{2-p} \Delta u^p(x,t)\, dt$

by Lemma 2 $\qquad \leq A_p(f^*(x))^{2-p} \int_0^\infty t\Delta F(x,t)\, dt$, (since $u(x,t) \leq f^*(x)$).

Let $I(x) = \int_0^\infty t\Delta F(x,t)\, dt \geq 0$. We have just proved that

$$(g(f)(x))^2 \leq A_p(f^*(x))^{2-p} I(x) .$$

On the other hand,

$$\int_G I(x)dx = \int_G F(x,0)dx - \int_G F(x,+\infty)dx \leq \int_G F(x,0)dx$$

$$= \int_G |f(x)|^p dx ,$$

by Lemma 3. Hence

$$\|g(f)\|_p^p = \int_G g(f)^p dx \leq A \int_G (f^*(x))^{(2-p)/p} I(x)^{p/2} dx$$

$$\leq A\left(\int_G |f^*(x)|^p dx\right)^{(2-p)/2} \left(\int_G I(x)\, dx\right)^{p/2}$$

(by Hölder's inequality), $\quad \leq A'\|f\|_p^{p(2-p)/2} \cdot \|f\|_p^{p\cdot p/2} = A\|f\|_p^p .$

Thus $\|g(f)\|_p^p \leq A\|f\|_p^p$, which completes part I of the p·oof. Note that the above arbument simplifies in the case $p = 2$, since then f^* does not enter the calculations, and Lemma 1 becomes unnecessary.

Part II: Using part I, we shall show that $\|g(f)\|_p \leq A_p\|f\|_p$ for $4 \leq p < +\infty$. From this and part I we deduce that $\|g(f)\|_p \leq A_p\|f\|_p$ for every p $(1 < p < +\infty)$. In fact, we have only to apply the M. Riesz or Marcinkiewicz interpolation theorem.

Suppose that f is a positive C^∞ function on G, p is a number ≥ 4, and q is the exponent conjugate to $p/2$. Thus $1 < q \leq 2$. Assume $\|f\|_p = 1$.

Now $\|g(f)\|_p^2 = \sup \int_G (g(f)(x))^2 \phi(x)dx$, where the supremum is taken over all positive C^∞ functions ϕ such that $\|\phi\|_q \leq 1$. Our plan is to prove

$$(*) \quad \int_G g(f)^2(x)\phi(x)dx \leq A_p\left[\int_G f(x)^2\phi(x)dx + \int_G f^*(x)g(f)(x)g(\phi)(x)dx\right]$$

From this inequality part II is not hard. The reason is that the left-hand side involves $g(f)^2$, while the right-hand side involves only $g(f)$. $f^*(x)$ and $g(\phi)$ are under control, the latter by part I and the fact that $\|\phi\|_q \leq 1$. So the size of $g(f)$ will be controlled.

The crucial step in proving (*) is to make the estimate

$$(**) \quad \int_G (g(f)(x))^2\phi(x)dx = \int_0^\infty \int_G t|\nabla u(x,t)|^2 \phi(x)dxdt$$

$$\leq A\int_0^\infty \int_G t|\nabla u(x,t)|^2\phi(x,t)dt$$

where $\phi(x,t)$ is the Poisson integral of ϕ. To prove (**) we will use a "subharmonicity" argument.

We remark first, and this is basic, that our Laplacian Δ is bi-invariant, and so commutes with the X_j. Thus any function of it also has this property, and so in particular for $T^t = e^{t\Delta}$, $t > 0$, and the Poisson semi-group. This commutitivity is obvious on the formal level and immediately justifiable on the subspace $\mathcal{E}$. The passage to general C^∞ functions is then by a routine limiting argument

Since $\{P^t\}$ is a semigroup, $u(x, s_1 + s_2) = P^{s_1}u(x, s_2)$; this shows that $X_ju(x, s_1 + s_2) = P^{s_1}X_ju(x, s_2)$ and $\frac{\partial u}{\partial s}(x, s_1 + s_2) = P^{s_1}\frac{\partial u}{\partial s}(x, s_2)$. Hence $X_ju(x, t) = P^{t/2}X_ju(x, t/2)$ and $\frac{\partial}{\partial t}u(x, t) = P^{t/2}\frac{\partial u}{\partial t}(x, t/2)$.

Next, note that since $|\nabla u|^2 = \Sigma_{ij}\, a_{ij}(X_i u)(X_j u) + (\frac{\partial u}{\partial t})^2$, we can, by a change of bases in the Lie algebra, assume that

$$|\nabla u|^2 = \sum_i (X_i u)^2 + (\frac{\partial}{\partial t} u)^2 .$$

So if we prove that $|X_j u(x,t)|^2 \le P^{t/2}(|X_j u(x,t/2)|^2)$ and similarly for $(\frac{\partial u}{\partial t})^2$, we shall have shown that $|\nabla u(x,t)|^2 \le P^{t/2}(|\nabla u(x,t/2)|^2)$. But $X_j u(x,t) = P^{t/2} X_j u(x,t) = \int_G p_{t/2}(xy^{-1}) X_j u(y,t/2) dy$ (where p_t denotes the Poisson kernel) $\le (\int_G p_{t/2}(xy^{-1}) |X_j u(y,t/2)|^2 dy)^{1/2}$ (Hölder's inequality) $= (P^{t/2}(X_j u)^2)^{1/2}$. Hence, $|\nabla u(x,t)|^2 \le P^{t/2}(|\nabla u(x,t/2)|^2)$ —which says, more or less, that the square of a (particular) harmonic function is subharmonic.

We can now prove (**). By the above,

$$\int_G g^2(f)(x)\phi(x)dx \le \int_0^\infty \int_G t P^{t/2}\,(\nabla u(x,t/2)|^2 P^{t/2}\phi(x) dx dt$$

(by the self-adjointness of $P^{t/2}$) $= \int_0^\infty \int_G t\,|\nabla u(x,t/2)|^2 \phi(x,t/2) dx\, dt$. If we make the change of variables $s = t/2$, we obtain (**).

Putting $|\nabla u(x,t)|^2 = \frac{1}{2}\Delta u^2(x,t)$ in (**), we obtain

$$\int_G (g(f)(x))^2 \phi(x) dx \le A \int_0^\infty \int_G t(\Delta u^2(x,t))\phi(x,t) dx dt .$$

It is tempting to integrate the right-hand side of this inequality by parts, since $\Delta\phi = 0$. We shall arrive at the result of such a computation in a slightly slicker way—namely, we shall use the elementary identity

$$\Delta(FG) = (\Delta F)\cdot G + F\cdot(\Delta G) + 2\sum_{ij} a_{ij}(X_i F)(X_j G) + 2(\frac{\partial F}{\partial t})(\frac{\partial G}{\partial t}) .$$

If we take $F = (u(x,t))^2$, $G = \phi$; and apply our identity to the last inequality, we obtain

$$(***)\qquad \int_G (g(f)(x))^2\phi(x)dx \le A\Bigg[\int_0^\infty\int_G t\Delta(FG)dxdt$$

$$-2\int_0^\infty\int_G\Bigg[\sum_{ij} a_{ij}X_j\ (u^2(x,t))X_i\phi(x,t)$$

$$+\left(\frac{\partial u^2}{\partial t}(x,t)\right)\left(\frac{\partial}{\partial t}\phi(x,t)\right)\Bigg]dx\,dt\Bigg]$$

$$\equiv A[\psi_1-\psi_2]\ .$$

We shall estimate ψ_1 and ψ_2 separately.

By Lemma 3, $\psi_1 = \int_G F(x)G(x)dx$ – positive $= \int_G f^2(x)\phi(x)dx$ – positive $\le \int_G f^2(x)\phi(x)dx$.

Since $X_ju^2 = 2uX_ju$ and $\frac{\partial}{\partial t}u^2 = 2u\frac{\partial}{\partial t}u$, we have

$$\psi_2 = 4\int_0^\infty\int_G tu(x,t)[\Sigma_{ij}\ a_{ij}(X_ju(x,t)(X_i\phi(x,t)) + (\frac{\partial u}{\partial t}(x,t))(\frac{\partial\phi}{\partial t}(x,t))]dxdt$$

$$\le 4\int_0^\infty\int_G tu(x,t)|\nabla u(x,t)|\ |\nabla\phi(x,t)|dxdt$$

$$(***)\qquad \le 4\int_G f^*(x)\int_0^\infty t|\nabla u(x,t)|\ |\nabla\phi(x,t)|dtdx$$

$$\le 4\int_G f^*(x)g(f)(x)g(\phi)(x)dx,\ \text{by Schwarz's inequality.}$$

Combining (***) with other estimates for ψ_1 and ψ_2, we obtain (finally!) the basic inequality (*).

Now we can prove part II. Applying Hölder's inequality to (*), we obtain

$$\int_G (g(f)(x))^2\phi(x)dx \le A_p[\|f\|_p^2 + \|f^*\|_p\ \|g(\phi)\|_q].$$

By part I, $\|g(\phi)\|_q \le A_q$ for $\|\phi\|_q \le 1$, so

$$\int_G (g(f)(x))^2\phi(x)dx \le A_p' [\,\|f\|_p^2 + \|f^*\|_p\,\|g(f)\|_p]$$

$$\le A_p'' [\,\|f\|_p^2 + \|f\|_p\,\|g(f)\|_p] \;.$$

(We have used Lemma 1 to deduce the last inequality.) Taking the sup over all ϕ (with $\|\phi\|_q \le 1$) yields $\|g(f)\|_p^2 \le A_p''[\|f\|_p^2 + \|f\|_p\|g(f)\|_p]$, which shows that $\|g(f)\|_p \le A_p'''\|f\|_p$. This completes part II.

We have just shown that $\|g(f)\|_p \le A_p\|f\|_p$ for $1 < p < +\infty$, which is half of Theorem 2. Using a standard duality argument, we can prove the converse inequality, and even a bit more. In fact, for $f \in L_p$, define the g_1*-function* for f by $g_1(f)(x) = (\int_0^\infty t\,|\frac{\partial}{\partial t}\, u(x,t)|^2\, dt)^{1/2}$. Obviously $g_1(f)(x) \le g(f)(x)$. We shall show that $\|f\|_p \le B_p\|g_1(f)\|_p$ if $\int_G f(x)dx = 0$.

To do so, we observe that $f \to g_1(f)$ is essentially an isometry on $L_2(G)$. More precisely, let $f \in L_2$ and let $c = \int_G f(x)dx$. Then

$$\text{(A)} \qquad \|f\|_2^2 = c^2 + 4\|g_1(f)\|_2^2 \;.$$

Formally, this identity is a consequence of the Parseval theorem for the complete orthonormal system ψ_k of eigenvectors of Δ. For if $f \sim \Sigma\, a_k\psi_k(x)$, then $u = P^t f \sim \Sigma_k\, e^{-(\mu_k)^{1/2}\cdot t} a_k\psi_k(x)$, so $\frac{\partial}{\partial t}\, u(x,t) \sim -\Sigma_{k\neq 0}(\mu_k)^{1/2}\, e^{-(\mu_k)^{1/2} t} a_k\psi_j(x)$. Hence

$$\int_G (\frac{\partial u}{\partial t}\,(x,t))^2 dx = \sum_{k\neq 0} \mu_k\, e^{-2(\mu_k)^{1/2} t} a_k^2 \;,$$

so

$$\int_G g_1(f)(x)^2 dx = \int_0^\infty t \int_G (\frac{\partial u}{\partial t}\,(x,t))^2 dx\, dt = \sum_{k\neq 0} a_k^2 \int_0^\infty \mu_k t e^{-2(\mu_k)^{1/2} t}\, dt$$

$$= \frac{1}{4} \sum_{k\neq 0} a_k^2 \;.$$

This shows that $4\|g_1(f)\|_2^2 + c^2 = c^2 + \Sigma_{k\neq 0}\, a_k^2 \equiv \|f\|^2$. Justification for these formal manipulations is routine and is left to the reader.

By polarizing identity (A) we obtain the equation

$$4\int_0^\infty \int_G t\frac{\partial}{\partial t}u_1 \cdot \frac{\partial}{\partial t}u_2\, dx\, dt + c_1c_2 = \int_G f_1(x)f_2(x)\, dx \ ,$$

where $u_i(x,t) = P^t u_i(x)$ and $c_i = \int_G f_1(x)dx$, valid for $f_1, f_2 \in L_2(G)$. If $f \in L_2 \cap L_p$ $(1 < p < +\infty)$ and $c \equiv \int_G f(x)dx = 0$, then by the above,

$$\|f\|_p = \sup_{\substack{f_2 \leq L_2 \cap L_q \\ \|f_2\|_q \leq 1}} \int_G f(x)f_2(x)dx$$

$$= 4\sup \int_G \int_0^\infty t\frac{\partial}{\partial t}u_1(x,t) \cdot \frac{\partial}{\partial t}u_2(x,t)dx\,dt$$

$$\leq 4 \sup \int_G g_1(f)g_1(f_2)dx$$

$$\text{(by Hölder,s inequality)} \leq \Bigg(4 \sup_{\substack{f_2 \in L_2 \cap L_q \\ \|f_2\|_q \leq 1}} \|g_1(f_2)\|_q\Bigg) \|g_1(f)\|_p$$

$$\leq B_p \|g_1(f)\|_p$$

by virtue of the inequality $\|g_1(f_2)\|_q \leq \|g(f_2)\|_q \leq Ap\|f_2\|_q$, which we which we already know. Hence $\|f\|_p \leq B_p\|g_1(f)\|_p$, and Theorem 2 is proved. QED.

The following observation is in order. The arguments just given are possible because of the following two properties of the Laplacian Δ, which incidentally are *not* shared by general Laplace-Beltrami operators. (1) Δ is a quadratic expression in first order differential operators. (2) Δ commutes with these first order operators. We shall return to this special situation in Chapter V.

Section 4. Applications: Riesz transforms, etc.

The following examples, chosen for their simplicity, illustrate the applications of the Littlewood-Paley inequalities to the Fourier analysis on the group G.

1. *Riesz Transforms.* Let $X_1 \cdots X_n$ be a base for the left-invariant vector fields on G. The j-th *Riesz transform* of f is given by the formal equation $R_j f = X_j(-\Delta)^{-1/2} f = (-\Delta)^{-1/2} X_j f$. This makes sense, because for $f = \Sigma_k a_k \psi_k(x) \in \mathfrak{S}$ we can define $(-\Delta)^{-1/2} f$ rigorously as

$$\sum_{k \neq 0} \frac{a_k}{(\mu_k)^{1/2}} \psi_k(x) .$$

As in the "classical" case $G = R^n$ (which strictly speaking, recall, doesn't fall into our general framework, since it isn't compact), the Riesz transforms may be used to prove miscellaneous L_p inequalities involving derivatives. For example, we can show that $\|X_i X_j f\|_p \leq A_p \|\Delta f\|_p$ $(1 < p < +\infty)$.

By using our Littlewood-Paley inequalities we can easily prove the basic property of Riesz transforms, namely that R_j is a bounded operator on L_p $(1 < p < +\infty)$. In fact, let $f = \Sigma a_k \psi_k(x)$ belong to $\mathfrak{S}$, and let $f_j = R_j f$. Consider the Poisson integrals u and u_j corresponding to f and f_j respectively. We have $\partial u/\partial t = -P^t((-\Delta)^{-1/2} f)$ since both expressions are equal to $-\Sigma_{k\neq 0}(\mu_k)^{1/2} e^{-\mu_k t} a_k \psi_k(x)$ —all the equation really says is that $-(-\Delta)^{1/2}$ is the infinitessimal generator of $\{P^t\}$. Applying R to both sides and noting that everything commutes we obtain

$$\frac{\partial u_j}{\partial t} = - X_j u(x,t) .$$

Therefore

$$g_1(R_j f)(x) = \left(\int_p^{\infty} t \left| \frac{\partial}{\partial t} u_j(x,t)\right|^2 dt\right)^{1/2} \leq K\left(\int_0^{\infty} t|\nabla u(x,t)|^2 dt\right)^{1/2}$$

$$= Kg(f) .$$

Since $\|g_1(R_j f)\|_p \approx \|R_j f\|_p$ and $\|g(f)\|_p \approx \|f\|_p$, we conclude that R_j is a bounded operator on $L_p(G)$.

Problems:

(1) (possibly not difficult). Show that the R_j are pseudo-differential operators on G. More particularly, show that the R_j are obtainable from the standard Euclidean singular integral operators by patching together. (For an introduction to the latter see e.g. Palais et al [5].) (2) Make a systematic study of the pseudo-differential operators on G which commute with the left action of G.

2. *Riesz Potentials.* The operator $(-\Delta)^{i\gamma}$ (γ real), which as usual is rigorously defined on functions in $\mathcal{E}$, is of interest in analysis. We shall prove

THEOREM 3: $\|(-\Delta)^{i\gamma} f\|_p \leq A_p(\gamma) \cdot \|f\|_p$.

It would be interesting to know whether $(-\Delta)^{i\gamma}$ is also a pseudo-differential operator. This problem is analogous to the same question for R_j.

To prove Theorem 3, we note that $\lambda^{i\gamma} = K\lambda \int_0^\infty e^{-\lambda s} \cdot s^{-i\gamma} ds$ for $\lambda > 0$. A function $m(\lambda)$ is said to be of *Laplace-transform type* if we can write $m(\lambda) = \lambda \int_0^\infty e^{-\lambda s} a(s) ds$, $\lambda > 0$, for some function $a(s)$ uniformly bounded on $(0, \infty)$ —say $|a(s)| \leq M$ (all $s \in (0,\infty)$). Thus Theorem 3 is a special case of

THEOREM 3′: *Let* $m(\lambda)$ *be a function of Laplace-transform type. Define the operator* T *on* $\mathcal{E}$ *by setting*

$$Tf = \sum_k m((\mu_k)^{1/2}) \cdot a_k \psi_k(x) \ \text{if} \ f = \sum_k a_k \psi_k(x) \in \mathcal{E} .$$

Then $\|Tf\|_p \leq A_p \|f\|_p$ $(1 < p < +\infty)$. *If we take* $m(\lambda) = \lambda^{2i\gamma}$ *in Theorem 3′ we obtain Theorem 3.*

Theorem 3′ corresponds to a weakened form of the classical Marcinkiewicz multiplier theorem for the n-torus T^n. The connection is that if

$m(\lambda)$ is of Laplace transform type, then $|\lambda^k m^{(k)}(\lambda)| \leq A_k$ for every $k \geq 0$, while a special case of the Marcinkiewicz multiplier theorem says that if $|\lambda^k m^{(k)}(\lambda)| \leq A_k$ for each $k \leq [(n+1)/2]$, then $m(|x|)$ is a multiplier on $L_p(T^n)$ $(1 < p < +\infty)$.

Proof of Theorem 3′: For a (reasonable) function f, define the Littlewood-Paley g_2*-function* of f by the equation

$$g_2(f)(x) = \left(\int_0^\infty t^3 \left| \frac{\partial^2}{\partial t^2} u(x,t) \right|^2 dt \right)^{1/2} .$$

We shall prove that $\|g_2(f)\|_p \approx \|f\|_p$ and that $g_1(Tf)(x) \leq K g_2(f)(x)$.

First of all, $\|f\|_p \leq B_p \|g_2(f)\|_p$ (if $\int_G f(x)dx = 0$) is quite easy. For

$$\left| \frac{\partial}{\partial t} P^t f \right| \leq \int_t^\infty \left| \left(\frac{\partial^2}{\partial \tau^2} P^\tau f \right) \right| d\tau = \int_t^\infty \left| \frac{\partial^2}{\partial \tau^2} P^\tau f \right| \tau \frac{d\tau}{\tau}$$

$$\leq \left(\int_t^\infty \left| \frac{\partial}{\partial \tau^2} P^\tau f \right|^2 \tau^2 \, d\tau \right)^{1/2} t^{-1/2}$$

by Hölder's inequality. Hence

$$(g_1(f)(x))^2 = \int_0^\infty t \left| \frac{\partial}{\partial t} P^t f(x) \right|^2 dt \leq \int_0^\infty \left(\int_t^\infty \left| \frac{\partial^2}{\partial \tau^2} P^\tau f(x) \right|^2 \tau^2 \, d\tau \right) dt$$

$$= \int_0^\infty \tau^3 \left| \frac{\partial^2}{\partial \tau^2} P^\tau f(x) \right|^2 d\tau = (g_2(f)(x))^2$$

by Fubini's theorem. Thus $g_1(f) \leq g_2(f)$, so we have $\|f\|_p \leq B_p \|g_2(f)\|_p$ by Theorem 2.

The converse inequality $\|g_2(f)\|_p \leq A_p \|f\|_p$ is more difficult, and we proceed with stealth.

We prove it here only for $p \geq 2$, and this suffices for our immediate purposes. The full case $1 < p < \infty$ is contained in the general treatment of Chapter IV.

Consider the Poisson kernel p_t defined by the identity $P^t f = p_t * f$. We note that

1. $t \int_G |\frac{\partial}{\partial t} p_t(x)| dx \leq C$ for all $t > 0$, as follows immediately from the argument given in 2. below.

2. For a reasonable function f on G, define $f^{**}(x) = \sup_{t>0} (t|\frac{\partial}{\partial t} p_t| * f)(x)$.

Then $\|f^{**}\|_p \leq A_p \|f\|_p$ $(1 < p < +\infty)$. This is proved in the same way as we proved the corresponding inequality for f^*. In fact, let K_t be the kernel for the heat semigroup (i.e., $T^t f = K_t * f$). By the subordination of P^t to T^t we have

$$p_t = \frac{1}{\sqrt{2\pi}} \int_0^\infty te^{-t^2/4s} K_s \cdot s^{-3/2}\, ds,$$

so that

$$t|\frac{\partial}{\partial t} p_t(x)| \leq \frac{1}{2\pi}\, t \int_0^\infty |\frac{\partial}{\partial t} te^{-t^2/4s}| K_s(x)\, s^{-3/2} ds\, .$$

The last inequality allows us to deduce $\|f^{**}\|_p \leq A_p \|f\|$ from the Hopf-Dunford-Schwartz ergodic theorem.

Now set

$$g_*(f)(x) = \left(\int_G \int_0^\infty t^2 |\frac{\partial}{\partial t} p_t(xy^{-1})|\, |\nabla u(y,t)|^2 dy\, dt \right)^{1/2}$$

(as usual $u(x,t) = P^t f(x)$). We shall prove that

(A) $$\|g_*(f)\|_p \leq A_p \|f\|_p \quad (2 < p < +\infty),$$

and

(B) $$g_2(f)(x) \leq C g_*(f)(x).$$

(A) and (B) together imply $\|g_2(f)\|_p \leq A_p \|f\|_p$ $(2 < p < +\infty)$.

To prove (A) we take arbitrary $\phi \geq 0$ a satisfying $\|\phi\|_q \leq 1$, where q is the exponent conjugate to p_2. By (2) above, $\|\phi^{**}\|_q \leq K$. So

$$\int_G (g_*(f)(x))^2 \phi(x)dx = \int_G \int_G \int_0^\infty t^2 \left|\frac{\partial}{\partial t} p_t(xy^{-1})\right| \, |\nabla u(y,t)|^2 \phi(x) dy\,dx\,dt$$

$$\leq \int_G \int_0^\infty t |\nabla u(y,t)|^2 \phi^{**}(y) dt\,dy$$

$$= \int_G (g(f)(y)^2 \phi^{**}(y)dy \leq \sup_{\|\psi\|_q \leq K} \int_G (g(f)(y))^2 \psi(y)dy$$

$$= K\|g(f)\|_p^2 \leq A_p \|f\|_p^2 \ .$$

Taking the sup over all $\phi \geq 0$ with $\|\phi\|_p \leq 1$, we obtain

$$\|g_*(f)\|_p^2 \leq A_p \|f\|_p^2 \ .$$

Statement (B) is even easier. If $t = t_1 + t_2$, then $P^t f = p_{t_1} * p_{t_2} * f$, so

$$\frac{\partial^2}{\partial t^2} P^t f = \left(\frac{\partial}{\partial t_1} p_{t_1}\right) * \left(\frac{\partial}{\partial t_2} p_{t_2}\right) * f \leq \left|\frac{\partial}{\partial t_1} p_{t_1}\right| * \frac{\partial P^{t_2}}{\partial t_2} f \ .$$

Taking $t_2 = t_2 = t/2$, we can say that

$$(g_2(f)(x))^2 = \int_0^\infty t^3 \left|\frac{\partial^2}{\partial t^2} p^t f(x)\right|^2 dt$$

$$\leq \int_0^\infty t^3 \left(\int_G \left|\left(\frac{\partial}{\partial t} p_t\right)_{t/2}(xy^{-1})\right| \, \left|\left(\frac{\partial}{\partial t} P^t f(y)\right)_{t/2}\right| dy \right)^2 dt$$

$$\leq \int_0^\infty t^3 \left(t^{-1} \int_G \left|\left(\frac{\partial}{\partial t} p_t\right)_{t/2}(xy^{-1})\right| \, \left|\left(\frac{\partial}{\partial t} P^t\right)_{t/2} f(y)\right|^2 dy \right) dt$$

$$\text{(by Schwarz's inequality)} \leq \int_G \int_0^\infty t^2 \cdot \left(\frac{\partial}{\partial t} p_t\right)_{t/2}(xy^{-1}) | \nabla u(y,t)|^2 dt\,dy$$

$$= \frac{1}{8} (g_*(f)(x))^2$$

by virtue of the change-of-variable $s = t/2$. This completes the proof of (B).

Finally, we have indeed shown that $\|g_2(f)\|_p \leq \|f\|_p$, at least for $p > 2$.

Remember why we have gone through this elaborate contortion? We are trying to prove that if $m(\lambda) = \lambda \int_0^\infty e^{-\lambda t} a(t)dt$ is of Laplace-transform type, then the operator T defined by

$$T(\sum_k a_k \psi_k(x)) = \sum_k m((\mu_k)^{1/2}) a_k \psi_k(x)$$

is bounded on $L_p(G)$.

Very well, then. Let $f \in \mathcal{E}$ have Poisson integral u and let $F \equiv Tf$. We claim that

$$F(x) = -\int_0^\infty \frac{\partial u}{\partial t}(x,t) a(t)dt \ .$$

For it suffices to check this identity for $f = \psi_k$; in that case

$$u(x,t) = e^{-(\mu_k)^{1/2} t} \psi_k(x) ,$$

so that

$$\int_0^\infty \frac{\partial u}{\partial t} a(t)dt = \int_0^\infty -(\mu_k)^{1/2} e^{-(\mu_k)^{1/2} t} \psi_k(x) a(t)dt$$

$$= - m((\mu_k)^{1/2}) \psi_k(x) = -T\psi_k(x) .$$

Applying the Poisson kernel to both sides of our identity yields

$$P^{t_1} F(x) = -\int_0^\infty P^{t_1}(\frac{\partial}{\partial t} P^t f(x)) a(t)dt$$

$$= -\int_0^\infty \frac{\partial}{\partial t} P^{t+t_1} f(x) a(t)dt.$$

Hence

$$\frac{\partial}{\partial t_1} P^{t_1} F(x) = -\int_0^\infty \frac{\partial^2}{\partial t^2} P^{t_1+t} f(x) a(t) dt .$$

So

$$\begin{aligned} \left|\frac{\partial}{\partial t_1} P^{t_1} F(x)\right| &\le \int_0^\infty \left|\frac{\partial^2}{\partial t^2} P^{t+t_1} f(x)\right| |a(t)| dt \\ &\le M \int_0^\infty \left|\frac{\partial^2}{\partial t^2} P^{t+t_1} f(x)\right| dt \\ &= M \int_{t_1}^\infty t \left|\frac{\partial^2}{\partial t^2} u(x,t)\right| \frac{dt}{t} . \end{aligned}$$

Repeating the arguments proving that $g_1(f) \le C g_2(f)$, we find that

$$\begin{aligned} (g_1(F)(x))^2 &= \int_0^\infty t_1 \left|\frac{\partial}{\partial t_1} P^{t_1} F(x)\right|^2 dt_1 \\ &\le C \int_0^\infty t^3 \left|\frac{\partial^2}{\partial t^2} u(x,t)\right|^2 dt = C(g_2(f)(x))^2 . \end{aligned}$$

Since $\|g_1(F)\|_p \approx \|F\|_p$ and $\|g_2(f)\|_p \approx \|f\|_p$ $(p > 2)$, it follows that T is a bounded operator on $L_p(G)$ for $p > 2$. The usual duality argument now shows that T is bounded on $L_p(G)$ for $p < 2$, and T is clearly bounded on $L_2(G)$. QED

Open problem: Obtain a strengthened multiplier theorem, which reduces to the Marcinkiewicz multiplier theorem in the classical case $G = T^n$, the n-torus.*

*See Appendix (1985).

BIBLIOGRAPHICAL COMMENTS FOR CHAPTER II

Section 1. Most of the results stated in Theorem 1 (the construction of heat-diffusion semi-group) could be read off from Hunt's paper [12].

Section 2, 3, and 4. The idea of this approach to the Littlewood-Paley theory originates in the author's course at Orsay [19], where it is carried out for $\mathbf{R}^n$. For some related ideas, see Gasper [41].

The original theory for $\mathbf{R}^n$ and its related realated real-variable approach owes much to the pioneering work of Besicovitch, Marcinkiewicz, and the paper of Calderón and Zygmund, "On the existence of certain singular integrals," Acta. Math., 88 (1952), 85-139. For the inequalities for the g-functions see the author's paper [18].

CHAPTER III

GENERAL SYMMETRIC DIFFUSION SEMI-GROUPS

Section 1. General setting.

We shall now develop the main theme of these lectures. Our purpose is to obtain such basic Fourier-analytic tools as the Hardy-Littlewood maximal function, the Littlewood-Paley inequalities, and conditions for L^p multipliers, in a very general setting. An appropriate context seems to be that of the symmetric diffusion semi-group. This notion is defined as follows.

$(\mathfrak{M}, dx)$ is some positive measure space. $\{T^t\}_{0<t<+\infty}$ is a family of operators, each T^t mapping functions on $\mathfrak{M}$ to functions on $\mathfrak{M}$. $\{T^t\}$ satisfies the semigroup axioms $T^{t_1+t_2} = T^{t_1}T^{t_2}$ and $T^0 = \text{identity}$. For any $f \in L_2(\mathfrak{M})$, all the $T^t f$ belong to $L_2(\mathfrak{M})$. and $\lim_{t\to 0} T^t f = f$ in L_2. Suppose also

(I) $\|T^t f\|_p \leq \|f\|_p \quad (1 \leq p \leq +\infty)$ (Contraction property)

(II) Each T^t is a self-adjoint operator on $L_2(\mathfrak{M})$ (Symmetry property)

(III) $T^t f \geq 0$ if $f \geq 0$ (Positivity property)

(IV) $T^t 1 = 1$ (Conservation property)

$\{T^t\}$ is called a *symmetric diffusion semigroup.* Sometimes we shall drop assumptions (III) and (IV), but (I) and (II) are absolutely essential for what follows.

Our introductory example is of course the Poisson semigroup. But symmetric diffusion semigroups occur often in analysis. We content ourselves here by noting some other examples:

1) Let $\mathfrak{M} = (a_1, a_2)$ be some finite, half-infinite, or infinite interval, and consider the second-order differential operator

$$Lf \equiv a(x)\frac{d^2f}{dx^2} + b(x)\frac{df}{dx} + c(x)f \ .$$

Assume that $a(x) > 0$, $c(x) \leq 0$, and that a, b, and c satisfy minimal smoothness conditions. It is not hard to find a non-trivial measure $q(x)dx$ on $\mathfrak{M}$, which makes L formally self-adjoint, i.e.,

$$\int_{\mathfrak{M}} f(x)\overline{Lg(x)}q(x)dx = \int_{\mathfrak{M}} Lf(x)\overline{g(x)}q(x)dx$$

for $f, g \in C_0^\infty(a_1, a_2)$. By imposing certain boundary conditions on functions to which we apply L, we can make L into a self-adjoint (unbounded) operator on $L_2(\mathfrak{M}, q(x)dx)$. One can then show that L is the infinitessimal generator of a semigroup $T^t = e^{tL}$, satisfying (I), (II), and (III). If $c(x) \equiv 0$ then (IV) also holds, with appropriate boundary conditions.

This example is very general. It includes the theory of expansions in Hermite polynomials, Legendre polynomials, trigonometric polynomials, and many other kinds of expansions. For details, the reader may consult Titschmarsh's book *Eigenfunction Expansions* [36] and a paper *Elementary Solutions for Certain Parabolic Partial Differential Equations* by H. P. McKean, Jr. [13]. For further discussion see Chapter V.

2) Parts of the theory of second-order operators extend to R^n. Let D be a domain in R^n with a reasonable boundary, or more generally a smooth manifold. We consider an elliptic second-order partial differential operator of the form

$$Lf = e(x)^{-1}\sum_{i=1}^{n}\frac{\partial}{\partial x_i}(a_{ij}(x)\frac{\partial}{\partial x_j}f) + c(x)f ,$$

where we assume $c(x) \leq 0$ and $e(x) > 0$. L is then formally self-adjoint with respect to the measure e(x)dx. As before, L subject to boundary conditions, generates a semigroup $\{T^t\}$ satisfying (I), (II), and (III); and if $c(x) \equiv 0$ then (IV) holds. For details, see Phillips, T.A.M.S. 1961, p. 62-84, [15].

3) An important special case is the Laplace-Beltrami operator on a compact Riemannian manifold. If the Riemannian manifold is a Lie group, we obtain the heat semigroup of Chapter II (and Chapter V for the non-compact case).

4) Given symmetric diffusion semigroup we can construct other symmetric diffusion semigroups, by "subordination." (For the facts concerning subordination, see Bochner [22] and Feller [10, Chapter 13].)

For now, we shall abandon examples, and study diffusion semigroups for their own sake.

Section 2. Analyticity of these semi-groups.

Our first result already shows that despite the misleadingly simple appearance of the assumptions I) - IV) some rather far-reaching implications may be drawn from them.

THEOREM 1. *Let* $(\mathfrak{M}, dx)$ *be a sigma-finite measure space, and suppose that the semigroup* $\{T^t\}$ *satisfies axioms* (I) *and* (II). *Let* $1 < p < +\infty$. *Then the map* $t \to T^t$ *has an "analytic continuation," i.e., it extends to an analytic* L_p*-operator-valued function* $t + i\tau \to T^{t+i\tau}$, *defined in the sector* S_p: $|\arg(t+i\tau)| < \frac{\pi}{2}(1 - |\frac{2}{p} - 1|)$.

The terms used in the statement above are explained by the following

DEFINITION: A function f mapping an open subset Ω of $\mathbf{C}$, into the complex Banach space B, is called *analytic* (or *holomorphic*) if for each continuous linear functional L on B, $z \to L(f(z))$ is an analytic function of $z \in \Omega$. A map $z \to T^z$ from Ω into the space of bounded operators on B is called analytic if, for each $b \in B$, the function $z \to T^z b$ is analytic from Ω to B.

Proof of the theorem: We shall first prove the result for $p = 2$. By the spectral theorem, the bounded self-adjoint operator T^1 has the form

$$T^1 f = \int_{-1}^{1} \lambda \, dE_\lambda(f) ,$$

where $\{E_\lambda\}$ is a resolution of the identity in $L_2(\mathfrak{M})$.

Since $T^1 = T^{1/2} T^{1/2} = (T^{1/2})^*(T^{1/2})$, it follows that the resolution $\{E_\lambda\}$ is concentrated on $[0, 1]$, i.e., $E_\lambda = 0$ if $\lambda < 0$. Hence

$$T^1 f = \int_0^1 \lambda \, d E_\lambda(f)$$

for $f \in L_2(\mathfrak{M})$. Repeated application of the uniqueness of positive definite square roots of operators (for a proof of uniqueness, see Riesz-Nagy [35], section 104) shows that

$$T^{1/2^k} f = \int_0^1 \lambda^{1/2^k} d E_\lambda(f) .$$

Therefore

$$T^{m/2^k} f = \int_0^1 \lambda^{m/2^k} d E_\lambda(f) ,$$

and by the strong continuity of the family $\{T^t\}$ we conclude finally that

$$T^t f = \int_0^1 \lambda^t d E_\lambda(f)$$

for every $f \in L_2(\mathfrak{M})$ and every $t > 0$.

It is now child's play to define $T^{t+i\tau}$ in the half-plane $t = \mathrm{Re}(t+i\tau) \geq 0$. We merely set

$$T^{t+i\tau} f = \int_0^1 \lambda^{t+i\tau} d E_\lambda(f) .$$

This concludes the proof of our theorem for the case $p = 2$. Note that $\|T^{t+i\tau}\|_2 \leq 1$ for $t \geq 0$.

The L_1 (L_∞) result is (vacuously) true since the "sector" S_1 (or S_∞) is just the positive real axis, where $\{T^t\}$ is already defined.

Finally, we shall interpolate between the L_2 result and the L_1 (or L_∞) result, to obtain the theorem for L_p, $1 < p < 2$, $(2 < p < +\infty)$. To perform the interpolation, we make use of a "convexity" theorem of the author, which generalizes the Riesz convexity theorem.

The extension of the Riesz convexity theorem is the following:

Let $\mathfrak{M}$ be a measure space and let $U(z)$ be a linear operator for each complex z in the strip $0 \leq \operatorname{Re} z \leq 1$. For every z in the strip, we suppose that $U(z)$ maps simple functions on $\mathfrak{M}$ (i.e., finite linear combinations of sets of finite measure) to measurable functions on $\mathfrak{M}$ which are integrable over all sets of finite measure. Suppose further that for any simple functions f and g on $\mathfrak{M}$, the function

$$z \to \int_{\mathfrak{M}} (U(z)f)\, g\, dx$$

is bounded and analytic in the region $0 < \operatorname{Re} z < 1$, and continuous in the closure. Finally, assume that $\|U(z)f\|_{p_0} \leq M_0\|f\|_{p_0}$ for $\operatorname{Re} z = 0$, and $\|U(z)f\|_{p_1} \leq M_1\|f\|_{p_1}$ for $\operatorname{Re} z = 1$, (f simple). Then if we set p between p_0 and p_1 by $1/p = (1-t)/p_0 + t/p_1$, we have $\|U(t)f\|_p \leq M_0^{1-t} M_1^t \|f\|_p$, ($f$ simple).

This theorem reduces to the Riesz-Thorin convexity theorem if we put $U(z)$ equal to the single operator U for every z.

Sketch of Proof of the Convexity Theorem. We shall make use of the so-called "Three-Lines Lemma": Let $\Phi(z)$ be a bounded analytic function in the strip $0 < \operatorname{Re} z < 1$, continuous in the closure. Suppose that $|\Phi(t)| \leq M_0$ for $\operatorname{Re} z = 0$ and $|\Phi(z)| \leq M_1$ for $\operatorname{Re} z = 1$. Then for $0 \leq t \leq 1$, $|\Phi(t)| \leq M_0^{1-t} M_1^t$.

Proof: The auxilliary function $\Psi(z) = (M_0/M_1)^z \Phi(z)$ is bounded, analytic in the strip, continuous in the closure, and satisfies $|\Psi(z)| \leq M_0$ for

$\operatorname{Re} z = 0$ or 1. By the Phragmen-Lindelöf maximum principle for the strip, $|\Psi(t)| \leq M_0$, i.e., $|\Phi(t)| \leq M_0^{1-t} M_1^t$. QED.

Now let f be a simple function with $\|f\|_p = 1$. To show that $\|U(t)f\|_p \leq M_0^{1-t} M_1^t$, we need only prove that

(*) $$\left| \int_{\mathfrak{M}} (U(t)f)\, g\, dx \right| \leq M_0^{1-t} M_1^t$$

for every simple function g for which $\|g\|_q \leq 1$, where q is the exponent conjugate to p. Inequality (*) follows from the three-lines lemma if we set up the right function Φ. In fact, we can write $f = F\ell$ and $g = G\mathfrak{g}$ where $F, G \geq 0$ and $|\ell(x)| = |\mathfrak{g}(x)| = 1$ for almost all x. Put

$$f_z \equiv F^{p\left(\frac{1-z}{p_0} + \frac{z}{p_1}\right)} \ell \quad \text{and} \quad g_z \equiv G^{q\left(\frac{1-z}{q_0} + \frac{z}{q_1}\right)} \mathfrak{g}$$

where q_0 and q_1 are the exponents conjugate to p_0 and p_1, respectively. Obviously $f_t = f$, $g_t = g$; $\|f_z\|_{p_0}, \|g_z\|_{q_0} \leq 1$ for $\operatorname{Re} z = 0$, and $\|f_z\|_{p_1}, \|g_z\|_{q_1} \leq 1$ for $\operatorname{Re} z = 1$. Hence the bounded analytic function

$$\Phi(z) \equiv \int_{\mathfrak{M}} (U(z) f_z)\, g_z\, dx$$

satisfies $|\Phi(z)| \leq$ (norm of $U(z)$ as an operator on L_{p_0}) $\leq M_0$ if $\operatorname{Re} z = 0$, and similarly $|\Phi(z)| \leq M_1$ if $\operatorname{Re} z = 1$. The three-lines lemma now shows that

$$|\Phi(t)| = \left| \int_{\mathfrak{M}} (U(t)f)\, g\, dx \right| \leq M_0^{1-t} M_1^t .$$

This completes the proof of (*). QED.

More detailed expositions of convexity theorems may be found in Stein [22], Dunford-Schwartz *Linear Operators* [9, Chapter 6, Section 10] and in Zygmund *Trigonometrical Series*, Vol. II, Chapter XII, [20]. Notice that the above argument has to be patched up for $p = +\infty$; the problem is not difficult.

Now we can return to semigroups and finish off the proof of Theorem 1. Recall that we have defined a family of operators $\{T^t\}$ for complex t in the right half-plane $\operatorname{Re} t \geq 0$, satisfying the properties

(a) $\|T^t f\|_2 \leq \|f\|_2$ (all t in the right half-plane).

(b) If $f, g \in L_2$ then $t \to \int_{\mathfrak{M}} (T^t f)\, g\, dx$ is an analytic function of t, bounded in $\operatorname{Re} t > 0$ and continuous in the closure.

(c) $\|T^t f\|_1 \leq \|f\|_1$ for $t > 0$ (real).

We want to interpolate between (a) and (c). So let $\eta > 0$ be arbitrary, let $-\pi/2 \leq \theta \leq \pi/2$, and define $U(z) f = T^{\eta e^{i\theta z}} f$. By (b) above $\{U(z)\}$ is an analytic family of operators, in the sense of the hypothesis of the above convexity theorem. Furthermore, (a) shows that $\|U(z) f\|_2 \leq \|f\|_2$ for $\operatorname{Re} z = 1$, and (c) shows that $\|U(z) f\|_1 \leq \|f\|_1 \leq \|f\|_1$ for $\operatorname{Re} z = 0$. Therefore, by the convexity theorem, $\|T^t f\|_p \leq \|f\|_p$ $(1 < p < 2)$ where $t = \eta e^{i\theta(2-\frac{2}{p})}$. Hence $\|T^t f\|_p \leq \|f\|_p$ whenever

$$|\arg t| \leq \frac{\pi}{2}\left(2 - \frac{2}{p}\right) = \frac{\pi}{2}\left(1 - \left|\frac{2}{p} - 1\right|\right),$$

for $\eta > 0$ and $\theta \in [-\frac{\pi}{2}, \frac{\pi}{2}]$ are arbitrary. We can prove an analogous result for $2 < p < +\infty$ by interpolation between L_2 and L_∞.

It remains to show that

$$t \to \int_{\mathfrak{M}} (T^t f)\, g\, dx$$

is bounded, analytic, and continuous in the closure of the sector S_p, for $f \in L_p$, $g \in L_q$. This follows from (b) when f and g are simple, so that letting $\{f_k\}$ (respectively $\{g_k\}$) be a sequence of simple functions tending in L_p (respectively L_q) to f, (respectively g), we find that $\int_{\mathfrak{M}} (T^t f) g\, dx$ is the uniform limit of the analytic functions

$$\int_{\mathfrak{M}} (T^t f_k)\, g_k\, dx \qquad \text{QED.}$$

As an application of Theorem 1, we can show that if $f \in L_p$ $(1 < p < +\infty)$ then for almost every x, $t \to T^t f(x)$ is a very smooth function on $(0, \infty)$. This is essential is we are to define a Littlewood-Paley g-function involving $\partial/\partial t\, T^t f(x)$.

LEMMA: *Let* $f \in L_p(\mathfrak{M})$, $1 < p < +\infty$. *For each* t, *we can redefine* $T^t f$ *on a set of measure zero, in such a manner that for every fixed* x, $T^t f(x)$ *is a real-analytic function of* $t \in (0, \infty)$.

Proof: By Theorem 1, the function $t \to T^t f \in L_p$ extends to an analytic L_p-valued function on the sector S_p. Now, our definition of an analytic function Φ from a region $\Omega \subseteq \mathbf{C}$ to a Banach space B was that for each continuous linear functional L on B, $z \to L(\Phi(z))$ is a complex-valued analytic function on Ω. But it is a standard fact of functional analysis that this definition of analyticity is equivalent to any other "reasonable" definition imaginable, for example.

Φ is continuous and satisfies Cauchy's integral formula;

$$\lim_{\Delta z \to 0} \frac{\Phi(z+\Delta z) - \Phi(z)}{\Delta z} \quad \text{exists in the norm topology on B.}$$

If $z_0 \in \Delta(z_0, \varepsilon) \subseteq \Omega$, then Φ has a power series expansion

$$\Phi(z) = \sum_{k=0}^{\infty} b_k (z - z_0)^k \quad \text{where } b_k \in B$$

valid in $\Delta(z_0, \varepsilon)$ and such that $\Sigma_{k=0}^{\infty} |b_k| r^k < +\infty$ for any $r < \varepsilon$.

We shall use the last definition of analyticity. So for any $t_0 > 0$, $T^t f = \Sigma_{k=0}^{\infty} f_k (t - t_0)^k$, for all t in some neighborhood $\Delta(t_0, 2\varepsilon)$, and $\Sigma_{k=0}^{\infty} \|f_k\|_p \varepsilon^k < +\infty$. Each f_k is an equivalence class of functions—pick a particular representative, which we also denote f_k. For $t \in \Delta(t_0, \varepsilon)$, we can modify $T^t f$ on a set of measure zero, in such a manner that

$$T^t f(x) = \sum_{k=0}^{\infty} f_k(x)(t - t_0)^k \qquad \text{(every x).}$$

This makes sense, because $\Sigma_{k=0}^{\infty} \varepsilon^k |f_k(x)| \leq +\infty$ for almost every x, as follows from $\Sigma_{k=0}^{\infty} \varepsilon^k \|f\|_p < +\infty$.

Now cover $(0, \infty)$ with countably many neighborhoods $\Delta(t_0, \varepsilon)$. The rest is quite tirival, and details are left to the reader. QED.

Section 3. The maximal theorem.

Our first main result is the following:

MAXIMAL THEOREM: *Let the semigroup* $\{T^t\}$ *satisfy* (I) *and* (II). *Then* (a) *the maximal function, defined by* $f^*(x) \equiv \sup_{t>0} |T^t f(x)|$, *satisfies the inequality*

$$\|f^*\|_p \leq A_p \|f\|_p , \qquad 1 < p \leq +\infty ;$$

(b) *if* $f \in L_p(\mathfrak{M})$, *then*

$$\lim_{t \to 0} T^t f(x) = f(x) \text{ a.e.} \qquad (1 < p < +\infty).$$

Proof: First of all, the theorem makes sense, by virtue of the last lemma. The plan of the proof is as follows:

1. We will prove the theorem for $p = 2$.
2. Next, we will give a very simple refinement of the theorem, valid for $p = 2$.
3. We will take the result of part 2 as information on $p = 2$, use the Hopf-Dunford-Schwartz ergodic theorem as information on $p = 1 + \varepsilon$.

Our theorem then follows from interpolation.

Step 1: This is essentially a Tauberian argument for we have (by the Hopf-Dunford-Schwartz ergodic theorem) strong bounds on the averages

$$\frac{1}{s} \int_0^s |T^t f(x)| \, dt ,$$

and we have some additional information—from this we can prove bounds for $T^t f(x)$. The extra information is provided by our good friend, the Littlewood-Paley function

$$g_1(f)(x) = \left(\int_0^\infty t \left| \frac{\partial}{\partial t} T^t f(x) \right|^2 dt \right)^{1/2} .$$

We now show that $\| g_1(f) \|_2 \leq c \| f \|_2$ where c is a universal constant. By the spectral theorem, we can write T^t in the form

$$T^t f = \int_0^\infty e^{-\lambda t} dE_\lambda f$$

where $\{E_\lambda\}$ is a resolution of the identity. (See the proof of Theorem 1 in Section 2. There we had in effect written $T^t = \int_0^1 \lambda^t \, d\tilde{E}(\lambda)$. We should then have $E(\lambda) = \tilde{E}(e^{-\lambda})$, $0 \leq \lambda \leq \infty$.) Hence

$$\frac{\partial}{\partial t} T^t f = - \int_{0+}^\infty \lambda e^{-\lambda t} dE_\lambda f$$

Ergo,

$$\int_{\mathfrak{M}} |g_1(f)(x)|^2 dx = \int_{\mathfrak{M}} \int_0^\infty t \left| \frac{\partial}{\partial t} T^t f(x) \right|^2 dt\, dx$$

$$= \int_0^\infty t \left\| \frac{\partial}{\partial t} T^t f \right\|_2^2 dt = \int_0^\infty t \int_{0+}^\infty \lambda^2 e^{-2\lambda t} (dE_\lambda(f), f) dt$$

(again by the spectral theorem)

$$= \int_{0+}^\infty \lambda^2 \int_0^\infty t e^{-2\lambda t} dt (dE_\lambda f, f) = \frac{1}{4} \int_{0+}^\infty (dE_\lambda f, f)$$

$$\leq \frac{1}{4} \| f \|_2^2 .$$

We are in position to prove step 1. Integrating by parts shows that

$$\int_0^s t\left(\frac{\partial}{\partial t} T^t f\right)dt = sT^s - \int_0^s T^t dt .$$

Therefore,

$$|T^t f(x)| \leq \left|\frac{1}{s} \int_0^s T^t f(x) dt\right| + \left|\frac{1}{s} \int_0^s t\left(\frac{\partial}{\partial t} T^t f(x) dt\right|$$

$$\leq \left|\frac{1}{s} \int_0^s T^t f(x) dt\right| + \left|\frac{1}{s}\left(\int_0^s t\,dt\right)^{1/2}\left(\int_0^s t\left|\frac{\partial}{\partial t} T^t f(x)\right|^2 dt\right)^{1/2}\right|$$

$$\leq Mf(x) + g_1(f),$$

where

$$Mf(x) \equiv \sup_{s>0} \left|\frac{1}{s} \int_0^s T^t f(x) dt\right| .$$

Hence $f^*(x) \leq Mf(x) + g_1(f)(x)$ so that

$$\|f^*\|_2 \leq \|Mf\|_2 + \|g_1(f)\|_2 \leq A\|f\|_2$$

by the Hopf-Dunford-Schwartz ergodic theorem and the L_2-boundedness of the g-function. This completes step 1 of the proof.

Step 2: For $f \in L_p$ and $k \geq 0$ define

$$f_k^*(x) = \sup_{t>0} \left|t^k \frac{\partial^k}{\partial t^k} T^t f(x)\right| .$$

f_0^* is the same as the maximal function f^*. We shall show that for each $k \geq 0$, $\|f_k^*\|_2 \leq A_k\|f\|_2$.

The proof copies the argument of step 1. In fact, integration by parts shows that

$$\int_0^t s\frac{\partial^2 T^2}{\partial s^2} ds = -2\int_0^t s\frac{\partial T^s}{\partial s} + t^2\frac{\partial T^t}{\partial t}$$

so that by Hölder's inequality,

$$\left|t\frac{\partial T^t}{\partial t} f(x)\right| \leq \sup 2\left|\frac{1}{t}\int_0^t s\frac{\partial T^s}{\partial s} f(x)ds\right| + g_2(f)(x),$$

where

$$g_2(f)(x) \equiv \left(\int_0^\infty s^3\left|\frac{\partial^2 T^s}{\partial s^2} f(x)\right|^2 ds\right)^{1/2}.$$

The first term on the left-hand side of this inequality has already been estimated in step 1, where we showed that

$$\left\|\sup_{t>0}\left|\frac{1}{t}\int_0^t s\frac{\partial T^s}{\partial s} f(x)ds\right|\right\|_2 \leq A\|f\|_2 .$$

So in order to complete step 2, for the maximal function f_1^*, we need only show that $\|g_2(f)\|_2 \leq A\|f\|_2$. This follows from the spectral theorem in a manner analogous to the proof in step 1 for $\|g_1(f)\|_2 \leq A\|f\|_2$.

Thus $\|f_1^*\|_2 \leq A\|f\|_2$. The general case $\|f_k^*\|_2 \leq A_k\|f\|_2$ follows by induction. We begin the inductive step by computing

$$\int_0^t s\frac{\partial^{k+1}}{\partial s^{k+1}} T^s f(x)ds$$

by parts. Details are left to the reader.

Step 3: So far, we know that

$$\left\|\sup_{t>0}\left|t^{-1}\int_0^t T^s f(x)ds\right|\right\|_p \leq A_p\|f\|_p \qquad \text{for } p > 1$$

(think of p very near to 1), and that

$$\left\|\sup_{t>0} t^k\frac{\partial^k}{\partial t^k} T^t f(x)\right\|_2 \leq A_k\|f\|_2 \qquad \text{for each } k \geq 0 .$$

Clearly, in order to interpolate between these inequalities, we should search for an analytic family of operators I^{α} which act on functions of one variable $t \in (0, \infty)$, such that

$$I^1(f)(t) = \int_0^t f(s)ds, \quad \text{and} \quad I^{-k}(f)(t) = \frac{\partial^k}{\partial t^k} f(t) .$$

The formal computation

$$I^n(f)(t) = \int_0^t \int_0^s \cdots \int_0^r f(r)\,dr \cdots ds\,dt = \frac{1}{(n-1)!} \int_0^t (t-s)^{n-1} f(s)ds$$

suggests a reasonable definition for I^{α}:

$$I^{\alpha}(f)(t) \equiv \frac{1}{\Gamma(\alpha)} \int_0^t (t-s)^{\alpha-1} f(s)ds , \qquad (\alpha \in \mathbb{C}) .$$

If $\operatorname{Re}\alpha > 0$ then the integral defining I^{α} converges absolutely for $f \in L_1(0, \infty)$, but if $\operatorname{Re}\alpha \leq 0$, the integral need not be defined. This is as it should be, for I^{-k} is supposed to be the k-th derivative. $I^{\alpha}(f)$ is called the α-th *fractional integral* of f.

To justify our definition of fractional integration, we shall prove the

LEMMA: *Let* f *be in* C_0^{∞} *on* $(0,\infty)$. *Then the function* $\alpha \to I^{\alpha}(f)$ *defined for* $\operatorname{Re}\alpha > 0$, *has an analytic continuation to all of* $\mathbb{C}$. *Furthermore, the functional equations* $I^{\alpha} I^{\beta} f = I^{\alpha+\beta} f$ *and* $I^0 f = f$ *hold.*

Proof: It is convenient to replace I^{α} by the "modified fractional integral" M^{α} defined by

$$M^{\alpha}(f)(t) \equiv t^{-\alpha} \frac{1}{\Gamma(\alpha)} \int_0^t (t-s)^{\alpha-1} f(s)ds = \frac{1}{\Gamma(\alpha)} \int_0^1 (1-s)^{\alpha-1} f(st)ds .$$

As we noted, this integral is absolutely convergent, and M^{α} is analytic, in $\operatorname{Re}\alpha > 0$. Write

$$M^{\alpha}(f)(t) = \frac{1}{\Gamma(\alpha)} \int_0^{1/2} (1-s)^{\alpha-1} f(st)\,ds + \frac{1}{\Gamma(\alpha)} \int_{1/2}^{1} (1-s)^{\alpha-1} f(st)\,ds$$

$$\equiv \text{(I)} + \text{(II)} .$$

There is no difficulty at all in continuing term (I) into the whole complex line $\alpha \in C$. Term (II) is not so simple, because of the singularity of $(1-s)^{\alpha-1}$ at $s = 1$ ($\mathrm{Re}\,\alpha \leq 0$). But we can evaluate (II) formally by integrating by parts:

$$(1) \qquad \text{(II)} = \frac{1}{\Gamma(\alpha+1)} \int_{1/2}^{1} (1-s)^{\alpha} \frac{d}{ds} f(st)\,ds + \frac{1}{\Gamma(\alpha+1)} \left(\frac{1}{2}\right)^{\alpha} f(\tfrac{1}{2}t) \quad .$$

If $\mathrm{Re}\,\alpha > -1$, this expression makes good sense—using it as a definition of (II), we obtain an analytic continuation of M^{α} into the region $\mathrm{Re}\,\alpha > -1$. If we integrate by parts (formally) once again (i.e., in (1)). we get a definition of M^{α} valid for $\mathrm{Re}\,\alpha > -2$. By successive integration by parts, we continue M^{α} into the region $\mathrm{Re}\,\alpha > -k$ for any $k > 0$. Thus M^{α} (and therefore also I^{α}) may be continued throughout C.

The semigroup relation $I^{\alpha} I^{\beta} = I^{\alpha+\beta}$ follows from a routine computation for $\mathrm{Re}\,\alpha$, $\mathrm{Re}\,\beta \geq 0$, and so holds for all $\alpha, \beta \in C$, by virtue of the analyticity of I^{α}. To show that I^0 is the identity, consider I^{α} ($\alpha > 0$), let $\alpha \to 0$, and apply a routine "approximation of the identity" argument. Details are left to the reader. QED.

In particular, since $I^k I^{-k} =$ identity, we have $I^{-k} f = (\partial^k/\partial t^k) f$.

The operators I^{α} and M^{α} are fraught with applications to our maximal functions. For, the inequalities between which we are trying to interpolate can be rephrased

$$\|\sup_{t>0} M^1(Tf)(t))\|_p \leq A_p \|f\|_p \qquad (1 < p < +\infty),$$

and

$$\|\sup_{t>0} M^{-k}(Tf)(t)\|_2 \leq A_k \|f\|_2 \ .$$

We are trying to show that

$$\| \sup_{t>0} M^0(Tf)(t)\|_p \le A_p \|f\|_p .$$

Two obstacles stand in the way of interpolation:

(a) To interpolate using the given family of operators

$$\mathfrak{M}_\alpha : f \to \sup_{t>0} |M^\alpha(T^0 f)(t)|,$$

we need inequalities not merely for $\mathfrak{M}_1$ and $\mathfrak{M}_{-k}$, but for $\mathfrak{M}_{1+iy}$ and $\mathfrak{M}_{-k+iy}$ for any y.

(b) The operators $\mathfrak{M}_\alpha$ are unfortunately *non-linear.*

Neither of these problems is very serious. Let us first take $\mathrm{Re}\,\alpha > 0$.

$$\begin{aligned}\mathfrak{M}_\alpha \phi &= \sup_{t>0} \frac{1}{|\Gamma(\alpha)|}\frac{1}{|t^\alpha|}\left|\int_0^t (t-s)^{\alpha-1}\phi(s)ds\right| \\ &\le \sup_{t>0} \left|\frac{\Gamma(\mathrm{Re}\,\alpha)}{\Gamma(\alpha)}\right| \cdot \frac{1}{|\Gamma(\mathrm{Re}\,\alpha)|} \cdot \frac{1}{|t^\alpha|}\int_0^t (t-s)^{\mathrm{Re}\,\alpha-1}|\phi(s)|ds \\ &= \left|\frac{\Gamma(\mathrm{Re}\,\alpha)}{\Gamma(\alpha)}\right| \mathfrak{M}_{\mathrm{Re}\,\alpha}|\phi| ,\end{aligned}$$

if ϕ is a decent function on $(0,\infty)$. If we use the fact that

$$|\Gamma(x+iy)| \sim e^{-\frac{\pi}{2}|y|} \cdot |y|^{(x-\frac{1}{2})} \cdot \sqrt{2\pi} \qquad \text{as } y \to \pm\infty$$

(see Titschmarsh, *Theory of Functions,* p 259) and apply the last inequality to the function $\phi(t) = T^t f(x)$, we obtain the inequality

$$\|\mathfrak{M}_\alpha\|_p \le K e^{\pi|\mathrm{Im}\,\alpha|} \|\mathfrak{M}_{\mathrm{Re}\,\alpha}\|_p$$

for any α and p, where the constant K is uniform in $\mathrm{Re}\,\alpha$, provided $\mathrm{Re}\,\alpha$ varies inside a bounded set.

Therefore, $\|\mathfrak{M}_{1+iy} f\|_p \le K_p e^{\pi|y|}\|f\|_p$ for any p $(1 < p < +\infty)$ and any real y; and $\|\mathfrak{M}_{-k+iy} f\|_2 \le A_k e^{\pi|y|}\|f\|_2$ for any positive integer k and any real y, similarly, if we use (1) and the integration by-parts that follows it.

To handle (b), we linearize our operators as follows: For any reasonable function $t(x)$, mapping our basic measure space $(\mathfrak{M}, dx)$ into $(0, \infty)$, define an operator $T_\alpha^{t(*)}$ on L_p by setting

$$T_\alpha^{t(*)}f(x) \equiv \frac{1}{\Gamma(\alpha)} t(x)^{-\alpha} \int_0^{t(x)} T^s f(x) ds .$$

Obviously $|T_\alpha^{t(*)}f(x)| \leq \mathfrak{M}_\alpha(f)(x)$ for any function. By our inequalities for $\mathfrak{M}_\alpha$,

(1) $$\| T_{1+iy}^{t(\cdot)} f \|_{p_0} \leq A_{p_0} e^{\pi|y|} \| f \|_{p_0}$$

and

(2) $$\| T_{-k+iy}^{t(\cdot)} f \|_2 \leq B_k e^{\pi|y|} \| f \|_2$$

where A_p and B_k are *independent* of the function $t(\cdot)$. Since the operators $T_\alpha^{t(\cdot)}$ are linear, we may immediately apply the convexity theorem (of the previous section) to inequalities (1) and (2). To do this, set

$$U(z) = e^{z^2} T_\alpha^{t(\cdot)}$$

where $\alpha = \alpha(z) = (1-z)(-k) + z$. The result is $\| T_0^{t(\cdot)} f \|_p \leq K \| f \|_p$ where p is determined by the equations

(3) $$\begin{cases} \dfrac{\theta}{p_0} + \dfrac{(1-\theta)}{2} = \dfrac{1}{p} \\ 1 \cdot \theta + (1-\theta)(-k) = 0 . \end{cases}$$

The "constant" K, whatever else it may depend on, is independent of $t(\cdot)$.

Any p $(1 < p < \infty)$ arises from equations (3) for some values of k and $p_0 > 1$, for we have only to pick k very large. Therefore, we have shown that for every p $(1 < p < \infty)$ for every measurable function $t(\cdot)$: $(\mathfrak{M}, dx) \to (0, \infty)$, the inequality

$$\|T_0^{t(\cdot)}f\|_p \equiv \|T^{t(\cdot)}f(x)\|_p \leq K_p\|f\|_p \tag{4}$$

is valid, with K_p independent of $t(\cdot)$.

Now we are (essentially) done. We have merely to pick our function $t(\cdot)$ in such a way that $|T^t f(x)| \geq \frac{1}{2} \sup_{t>0} |T^t f(x)|$ for each x. By inequality (4), $\|\sup_{t>0} |T^t f(x)|\|_p \leq 2K_p\|f\|_p$, In other words,

$$\|f^*\|_p \leq A_p\|f\|_p \qquad (1 < p < \infty),$$

which is exactly what we wanted to prove.

It remains only to show that $\lim_{t \to 0+} T^t f = f$ almost everywhere ($f \in L_p$, $1 < p < +\infty$). As always, the almost-everywhere theorem is an easy consequence of the L_p-boundedness of the maximal function.

First let $f \in L_2$. Since $T^t f(x)$ is a real-analytic function of $t \in (0, \infty)$ for almost all x, we have $\lim_{t \to 0+} T^t(T^s f) = T^s f$ almost everywhere, for each $s > 0$. Hence

$$\begin{aligned}\limsup_{t \to 0+} |T^t f(x) - f(x)| &\leq \limsup_{t \to 0+} |T^t(f - T^s f)(x)| \\ &\quad + \limsup_{t \to 0+} |T^t(T^s f)(x) - T^s f(x)| + |T^s f(x) - f(x)| \\ &\leq \sup_{t>0} |T^t(f - T^s f)(x)| + |T^s f(x) - f(x)| \\ &= (f - T^s f)^*(x) + |T^s f(x) - f(x)| .\end{aligned}$$

So

$$\|\limsup_{t>0} |T^t f(x) - f(x)|\|_2 \leq 2\|(f - T^s f)^*\|_2 \leq K\|f - T^s f\|_2$$

(by the L_2-boundedness of the maximal function) $\to 0$ as $s \to 0+$, by strong continuity of $\{T^s\}$ on L_2. This proves that

$$\limsup_{t \to 0+} |T^t f(x) - f(x)| = 0 \text{ a.e.},$$

in other words $T^t f \to f$ a.e. as $t \to 0+$, for $f \in L_2(\mathfrak{M})$.

Now let f belong to $L_p(\mathfrak{M})$ $(1 < p < +\infty)$, and suppose $\varepsilon > 0$ is given. We can find a function $g \in L_2 \cap L_p$ such that $\|f-g\|_p < \varepsilon$. Using the same trick as before, we write

$$\limsup_{t \to 0+} |T^t f(x) - f(x)| \leq |T^t f(x) - T^t g(x)|$$

$$+ \limsup_{t \to 0+} |T^t g(x) - g(x)| + |g(x) - f(x)|$$

$$\leq \sup_{t > 0} |T^t (f-g)(x)| + |(f-g)(x)| \text{ (since } T^t g(x) \to g(x) \text{ a.e. as } t \to 0^+),$$

$$= (f-g)^*(x) + (f-g)(x) .$$

So by L_p-boundedness of the maximal function, we have

$$\| \limsup_{t \to 0+} |T^t f(\ ) - f(\ )| \|_p \leq K \|f-g\|_p \leq K\varepsilon .$$

Letting $\varepsilon \to 0$, we get $\limsup_{t \to 0+} |T^t f(x) - f(x)| = 0$ a.e., which means that $T^t f \to f$ a.e. QED.

Section 4. A digression: L_2 theorems.

Before we continue our development of Littlewood-Paley theory, we shall pursue a digression. The above maximal theorem and ergodic theorem were posed for semigroups $\{T^t\}$ which were contractions $(\|T^t\|_p \leq 1)$ for *all* p $(1 \leq p \leq +\infty)$. It is interesting to consider purely L_2-variants of the maximal and ergodic theorems.

THEOREM 3: *Let* P *be a self-adjoint operator on* $L_2(\mathfrak{M})$, *satisfying* $\|P\|_2 \leq 1$ *and the positivity condition* $Pf \geq 0$ *if* $f \geq 0$. *Then*

$$\| \sup_n |P^n f(x)| \|_2 \leq C \|f\|_2$$

for any $f \in L_2(\mathfrak{M})$, *where* C *is a universal constant.*

REMARK: Is the theorem still true if we drop the positivity assumption? *A priori* the answer would seem to be no, since then the hypotheses would be purely Hilbert space (i.e., unitarily invariant) while notions like almost everywhere convergence are not unitarily invariant. In fact Burkholder has pointed out that the answer is indeed no, using consideration of his "semi-Gaussian subspaces," see [33].

That the condition of self-adjointness cannot be much modified is evident if we take instead of P an appropriate unitary operator. For example if $U f(x) = f(x+\theta)$ is a shift by an irrational θ on the circle $T^1 = R^1/Z$, then it is easy to construct an $f \in L_2(T^1)$, so that

$$\sup_{n \geq 0} |U^n f(x)| = \infty \quad \text{everywhere.}$$

Proof of the theorem: Define $A(n) = (n+1)^{-1}\Sigma_{k=0}^{n} P^k$. We shall first prove an L_2-maximal theorem for $Mf(x) \equiv \sup_n A(n)f(x)$ —then we shall prove our maximal theorem by using the same sort of g-function argument which served us so well in the maximal theorem for $\{T^t\}$.

If $f \geq 0$ then

$$A(n)A(m)f \leq 2[A(2n)f + A(2m)f] \tag{1}$$

To prove this, suppose $n \leq m$. Then

$$A(m)A(n)f = \frac{1}{(m+1)(n+1)} \sum_{k=0}^{n} P^k \sum_{j=0}^{m} P^j f = \frac{1}{(m+1)(n+1)} \sum_{\ell=0}^{m+n} \mu_\ell P^\ell f$$

where μ_ℓ is the number of ways in which ℓ can be written as $k + j$ with $k \leq n$, $j \leq m$. Obviously $\mu_\ell \leq n+1$ for any ℓ. So

$$A(m)A(n)f \leq \frac{1}{(m+1)(n+1)} \sum_{\ell=0}^{m+n} (n+1)P^\ell f = \frac{1}{(m+1)} \sum_{\ell=0}^{m+n} P^\ell f$$

$$\leq \frac{1}{m+1} \sum_{\ell=0}^{2m} P^\ell f \leq 2A(2m)f \leq 2[A(2m)f + A(2n)f].$$

The situation is analogous if $m \leq n$.

Now suppose that $\|P\|_2 = 1-\delta < 1$. Then $\|P^n\|_2 \leq (1-\delta)^n$. So

$$\| \sup_{n \geq 0} |P^n f(\cdot)| \|_2^2 \leq \left\| \left(\sum_{n=0}^{\infty} |P^n f(\cdot)|^2 \right)^{1/2} \right\|_2^2 = \sum_{n=0}^{\infty} \|P^n f\|_2^2$$

$$\leq \|f\|_2^2 \sum_{n=0}^{\infty} (1-\delta)^{2n} = A_\delta^2 \|f\|_2^2 .$$

It follows easily that $\|Mf\|_2 \leq A_\delta \|f\|_2$ for $f \in L_2$. Of course $A_\delta \to +\infty$ as $\delta \to 0$. But we shall show, using inequality (1), that $\|Mf\|_2 \leq C\|f\|_2$ where C is *independent of* δ.

Let C be the smallest constant such that $\|Mf\|_2 \leq C\|f\|_2$. As in the proof of the maximal theorem for semigroups, there is a function $n(x)$ defined on the measure space $\mathfrak{M}$, such that the operator $B: f \to A(n(x))f(x)$ on L_2 has norm $\geq C(1-\varepsilon)$ ($\varepsilon > 0$ is prespecified). B is a bounded operator on L_2 and $f \geq 0$ implies $Bf \geq 0$, but B is not necessarily self-adjoint. Define another operator B_2 on $L_2(\mathfrak{M})$ by setting $B_2(f)(x) = A(2n(x))f(x)$.

By inequality (1), $A(n(x))A(m) \leq 2[A(2n(x)) + A(2m)]$ in the obvious sense. In other words $BA(m)f \leq 2[B_2 + A(2m)]f$ for $f \geq 0$. We cannot pull a similar trick on $A(m)$, because it is being operated on by B. To overcome this difficulty, we pass to the adjoint. Thus

$$A(m)B^*f \leq 2[B_2^* f + A(2m)f] .$$

Taking $m = n(x)$, we obtain $BB^*f \leq 2[B_2^* + B_2]f$ for $f \geq 0$. The operators BB^* and $B_2^* + B_2$ map positive functions to positive functions, so $\|BB^*\|_2 \leq 2\|B_2^*\|_2 + 2\|B_2\|_2$. But by definition of B, B_2, and C, $\|BB^*\|_2 = \|B\|_2^2 \geq C^2(1-\varepsilon)^2$ and $\|B_2\|_2 \leq C$. So $\|B_2^*\| \leq C$ and $C^2(1-\varepsilon)^2 \leq 4C$. Since $\varepsilon > 0$ is arbitrary, $C \leq 4$.

We have shown that if $\|P\|_2 < 1$, then

$$\text{(2)} \qquad \left\| \sup_n \left| \frac{1}{(n+1)} \sum_{j=0}^{n} P^j f(\cdot) \right| \right\|_2 \leq 4\|f\|_2 .$$

But this inequality also holds if $\|P\|_2 = 1$, for we may apply (2) to the operator $P_\delta = (1-\delta)P$ and let $\delta \to 1-$. Details are left to the reader.

Now that we have a maximal inequality for the averages $A(n)$, we can easily prove our theorem.

Summation by parts shows that

$$Pf^n - \frac{1}{(n+1)} \sum_{k=0}^{n} P^k f = \frac{1}{(n+1)} \sum_{k=1}^{n} k[P^k - P^{k-1}]f$$

which has absolute value less then or equal to

$$\left(\sum_{k=1}^{\infty} k|P^k f(x) - P^{k-1} f(x)|^2 \right)^{1/2} \equiv g_1(f)(x) .$$

So to prove our theorem we need only show that $\|g_1(f)\|_2 \le C\|f\|_2$ for $f \in L_2$.

Suppose that P is positive definite in the Hilbert space sense $(Pf, f) \ge 0$. By the spectral theorem, we can write $P = \int_0^1 \lambda\, dE(\lambda)$. Hence

$$\|g_1(f)\|_2^2 = \sum_k k\|(P^k - P^{k-1})f\|_2^2 = \sum_k k \int_0^1 (\lambda^k - \lambda^{k-1})^2 (dE(\lambda)f, f)$$

$$= \int_0^1 [\sum_k k(\lambda^k - \lambda^{k-1})^2](dE(\lambda)f, f)$$

$$\le C \int_0^1 (dE(\lambda)f, f) \le C\|f\|_2^2 .$$

Thus, the maximal theorem $\|\sup_{n\ge 0}|P^n f(\cdot)|\,\|_2 \le C\|f\|_2$ is proved if we assume, in addition to the hypothesis, that P is positive-definite in the Hilbert space sense.

If P satisfies the hypothesis of the theorem but is not positive-definite, we apply what we already know to the operator P^2, which *is* positive-definite, to obtain $\|\sup_{n\ge 0}|P^{2n} f(\cdot)|\,\|_2 \le C\|f\|_2$. But

$$\sup_{n\ge 0}|P^n f(x)| \le \sup_{n\ge 0}|P^{2n}(Pf)(x)| + \sup_{n\ge 0}|P^{2n}(Pf)(x)|$$

so we have $\| \sup_{n \geq 0} |P^n f(\cdot)| \|_2 \leq C\|f\|_2 + C\|Pf\|_2 \leq C'\|f\|_2$. QED.

COROLLARY: $\lim_{n\to\infty} P^n f$ *exists almost everywhere, for every* $f \in L_2$, *if* P *is positive-definite in the Hilbert space sense.*

Proof: By the maximal theorem just proved, we need only show that $\lim_{n\to\infty} P^n f$ exists a.e. for f in a dense subset of $L_2(\mathfrak{M})$. (The argument proving a.e. convergence from a maximal theorem was already given twice at the end of the proof of the $\{T^t\}$ maximal theorem, and will not be repeated.) Consider f of the form $f = (E(1) - E(1-))g + E(1-\varepsilon)g$ for some $\varepsilon > 0$, $g \in L_2$ ($E(\cdot)$ denotes the resolution of the identity for P). Since $(E(1)-E(1-))g + E(1-\varepsilon)g \to g$ in $L_2(\mathfrak{M})$ as $\varepsilon \to 0^+$, the set of such f's is dense in L_2.

Let $f = (E(1) - E(1-))g + E(1-\varepsilon)g \equiv h_1 + h_2$. $P^n h_1 = h_1$, so $P^n h_1$ converges almost everywhere as $n \to +\infty$. On the other hand,

$$P^n h_2 = P^n E(1-\varepsilon)g = \int_0^{1-\varepsilon} \lambda^n dE(\lambda)g ,$$

so $\|P^n h_2\|_2 \leq C(1-\varepsilon)^n \|g\|_2$. Therefore

$$\sum_{n=0}^{\infty} \int_{\mathfrak{M}} |P^n h_2(x)|^2 dx \leq C^2 \sum_{n=0}^{\infty} (1-\varepsilon)^{2n} \|g\|_2^2 < +\infty ,$$

so

$$\int_{\mathfrak{M}} \left(\sum_{n=0}^{\infty} |P^n h_2(x)|^2 \right) dx < +\infty .$$

Hence $\sum_{n=0}^{\infty} |P^n h_2(x)|^2 < +\infty$ for almost all x, so $P^n h_2 \to 0$ almost everywhere as $n \to +\infty$.

Finally, then $P^n f = P^n h_1 + P^n h_2$ converges almost everywhere as $n \to +\infty$ for all f in a dense subset of $L_2(\mathfrak{M})$. QED.

The arguments used in the above theorem originate in the papers of Kolmogoroff-Selivestroff-Plessner in 1928, in which the authors prove the almost everywhere convergence of a Fourier series $\Sigma_{n=-\infty}^{\infty} a_n e^{in\theta}$ for which $\Sigma_{n=-\infty}^{\infty} |a_n|^2 \log|n| < +\infty$. Paley [29], Bochner (in his book *Fourier Analysis and Probability* 1955) [32] and E. Stein [31] successively widened the scope of these ideas.

The following two theorems are proved by essentially the same technique as Theorem 3.

THEOREM 4: (L_2 ergodic theorem). *Let* U *be a unitary operator on* $L_2(\mathfrak{M})$ *such that* $f \geq 0$ *implies* $Uf \geq 0$. *Define*

$$A_n^+(f) = \frac{1}{(n+1)} \sum_{j=0}^{n} U^j f$$

for $f \in L_2$. *Then* $\lim_{n\to\infty} A_n^+(f)$ *exists almost everywhere.*

Sketch of Proof: Let $A_n(f) \equiv \frac{1}{2n+1} \Sigma_{j=-n}^{n} U^j f$. The operators A_n are self-adjoint, and satisfy $A(n)A(m)f \leq 2[A(2n) + A(2m)]f$ if $f \geq 0$. Proceeding as in the proof of Theorem 3, we obtain the maximal inequality $\|\sup_{n\geq 0} A(n)f(\cdot)\|_2 \leq C\|f\|_2$. But for $f \geq 0$, $A_n^+ f \leq 2A_n f$, so the maximal inequality is proved. Almost everywhere convergence follows easily.

THEOREM5: (Martingale Convergence Theorem – L_2 variant). *Let* $E_1, E_2, \ldots$ *be the orthogonal projection operators to subspaces* $X_1 \subseteq X_2 \subseteq \ldots$ *of* $L_2(\mathfrak{M})$. *If* $f \geq 0$ *implies all* $E_j f \geq 0$, *then* $\lim_{n\to\infty} E_n f$ *exists almost everywhere, for each* $f \in L_2(\mathfrak{M})$.

Sketch of Proof: Use the inequality $E_m E_n f \leq E_m f + E_n f$ (if $f \geq 0$). This inequality is vlaid because $E_m E_n = E_{\min(m,n)}$. This implies that $\|\sup_n E_n f\|_2 \leq 2\|f\|_2$.

Some examples of applications of theorems 4 and 5 are in order. The typical operator U of Theorem 4 arises as follows: Let $(\mathfrak{M}, dx)$ be the interval [0, 1] with Lebesgue measure, and let $\phi : [0,1] \to [0,1]$ be some "reasonable" function. Then we define Uf by setting

$$Uf(x) = \left(\frac{d\phi}{dx}\right)^{1/2} f(\phi(x)) .$$

This example and its trivial generalizations include the classical situation in which U is induced by a measure-preserving transformation on $(\mathfrak{M}, dx)$.

Theorem 5 is much more pertinent for our purposes. To see its implications, consider the case $(\mathfrak{M}, dx) = ([0,1]$, Lebesgue measure), and set E_n equal to a "conditional expectation" operator:

$$E_n(f)(x) = \frac{1}{2^n}\int_{k/2^n}^{(k+1)/2^n} f(y)dy \text{ if } x \in \left[\frac{k}{2^n}, \frac{k+1}{2^n}\right] .$$

For this example, Theorem 5 amounts *in effect* to the classical differentiation theorem

$$f(x) = \lim_{y\to 0} \frac{1}{y}\int_0^y f(x+t)dt \qquad (f \in L_2) .$$

As the next step in our development of Littlewood-Paley theory, we shall analyse sequences of operators $\{E_n\}$, possessing the same essential properties as the above example (which we shall call the *dyadic interval* example).

BIBLIOGRAPHICAL COMMENTS FOR CHAPTER III

Sections 2 and 3. The Maximal theorem goes back to the author's paper [31]. Theorem 1 was implicit in that approach but has not been published before. A later proof of the maximal theorem was given by Rota [30]. but it requires properties (III) and (IV) in addition to (I) and (II). The required convexity for analytic family of operators appears in [22]. See also [25].
Section 4. An outline of the proof of Theorem 3 appears in the paper [31]. Theorems 4 and 5 are in the same spirit but were unpublished.

CHAPTER IV

THE GENERAL LITTLEWOOD-PALEY THEORY

Section 1. Conditional Expectation and Martingales

Let (M, dx) be a sigma-finite measure space, and let $\mathfrak{M}$ denote the family of measurable subsets of M. Suppose that $\mathfrak{F} \subseteq \mathfrak{M}$ is some smaller sigma-field, and assume (for technical reasons) that the "restricted" measure space $(M, \mathfrak{F}, dx)$ is also sigma-finite. We shall define an operator E which maps the measurable function f on $(M, \mathfrak{M}, dx)$ to its *conditional expectation* $E(f \mid \mathfrak{F})$.

To clarify the meaning of E, we first consider a simple example. Suppose that M is the disjoint union of a sequence $S_1, S_2, \ldots$ of measurable sets, where $0 < m(S_j) < +\infty$ for each j. The S_j's generate a sigma-field $\mathfrak{F} \subseteq \mathfrak{M}$. We shall call $\mathfrak{F}$ a *special* subfield of $\mathfrak{M}$.

Conditional expectations relative to special subfields of $\mathfrak{M}$ are easy to define. If (say)) $f \in L_1(M, \mathfrak{M}, dx)$ and $\mathfrak{F}$ is the special subfield of $\mathfrak{M}$, generated by $S_1, S_2, \ldots$, then we set

$$E_{\mathfrak{F}} f(x) = \frac{1}{m(S_j)} \int_{S_j} f(y)dy \quad \text{for } x \in S_j ,$$

and we call $E_{\mathfrak{F}}(f)$ the conditional expectation of f relative to the field $\mathfrak{F}$. The operators E_n defined above to illustrate Theorem 5 are obviously conditional expectation operators.

Note the fundamental property of $E_{\mathfrak{F}}$: $E_{\mathfrak{F}}(f)$ is $\mathfrak{F}$-measurable.

Now suppose that $\mathfrak{F} \subseteq \mathfrak{M}$ is any (sigma-finite) subfield. If f is a locally integrable function on $(M, \mathfrak{M}, dx)$ then there is an induced absolutely continuous measure λ on $\mathfrak{F}$ defined by $\lambda(A) = \int_A f(x)dx$, for $A \in \mathfrak{F}$. (The point is that we are restricting our attention to $A \in \mathfrak{F}$.) By the Radon-Nikodym theorem, there is an essentially unique $\mathfrak{F}$-measurable function g on M, such that $\lambda(A) = \int_A g(x)dx$ for $A \in \mathfrak{F}$.

In other words, for every f on M, we can find a unique $\mathfrak{F}$*-measurable* function g on M, such that $\int_A f(x)dx = \int_A g(x)dx$ for every $A \in \mathfrak{F}$. g is called the *conditional expectation* of f relative to $\mathfrak{F}$, and is written $g = E(f|\mathfrak{F})$.

We collect a few trivial properties of conditional expectations in the following list:

(1) $E(f|\mathfrak{F}) \geq 0$ if $f \geq 0$.

(2) If f is $\mathfrak{F}$-measurable, then $E(f|\mathfrak{F}) = f$. In particular $E(1|\mathfrak{F}) = 1$.

(3) $E(f+g|\mathfrak{F}) = E(f|\mathfrak{F}) + E(g|\mathfrak{F})$ and $E(\alpha f|\mathfrak{F}) = \alpha E(f|\mathfrak{F})$.

(4) $E(E(f|\mathfrak{F})|\mathfrak{F}) = E(f|\mathfrak{F})$

(5) If f and g are L_2 functions on M, then

$$\int_M E(f)\, g\, dx = \int_M f E(g) dx$$

(6) $\|E(f|\mathfrak{F})\|_p \leq \|f\|_p \quad (1 \leq p \leq +\infty)$

(7) If (say) g is bounded and $\mathfrak{F}$-measurable, then $E(gf|\mathfrak{F}) = g E(f|\mathfrak{F})$.

(8) If $\mathfrak{G} \subseteq \mathfrak{F}$ is another sigma-field, then $E(E(f|\mathfrak{F})|\mathfrak{G}) = E(f|\mathfrak{G})$.

Proof: These properties follow immediately from the definition of conditional expectation, but we still agree to writing out a proof for (5) and (6).

(5) $$\int_M E(f) \cdot g\, dx = \int_M E(E(f) \cdot g) dx = \int_M E(f) \cdot E(g) dx \text{ by (7),}$$

and

$$\int_M f \cdot E(g) dx = \int_M E(f \cdot E(g)) dx = \int_M E(f) \cdot E(g) dx \text{ by (7)}$$

(6): Since $E(f|\mathcal{F})$ is $\mathcal{F}$-measurable, we have (with $\frac{1}{p} + \frac{1}{q} = 1$)

$$\|E(f|\mathcal{F})\|_p = \sup\{ \textstyle\int_m Ef(x)g(x)dx \mid g \text{ is measurable in } \mathcal{F}, \text{ and } \|g\|_q \leq 1\}$$

$$= \sup\{ \textstyle\int_M f(x)g(x)dx \mid g \text{ is measurable in } \mathcal{F} \text{ and } \|g\|_q \leq 1\} \text{ (by (7))}.$$

$$\leq \sup\{ \textstyle\int_M f(x)g(x)dx \mid g \text{ is measurable in } \mathfrak{M} \text{ and } \|g\|_q \leq 1\} = \|f\|_p .$$

QED.

For the next idea I shall require an increasing sequence $\mathcal{F}_1 \subseteq \mathcal{F}_2 \subseteq \ldots$ of sigma-fields contained in $\mathfrak{M}$. If f is a reasonable function on M, let $E_n(f) = E(f|\mathcal{F}_n)$, and define $f^*(x) = \sup_{n\geq 1}|E_n(f)(x)|$.

THEOREM 6 ("Martingale Maximal Theorem"):

(a) $m\{x| f^*(x) > a\} \leq \frac{A}{a}\|f\|_1$, i.e., $f \to f^*$ has weak-type (1, 1)

(b) $\|f^*\|_p \leq A_p\|f\|_p \quad (1 < p \leq +\infty)$.

If $M = [0, 1]$ and $\mathcal{F}_n$ is the special field generated by the sets

$$\left[\frac{k}{2^n}, \frac{k+1}{2^n} \right], \quad k = 0, 1, \ldots, 2^n-1$$

then Theorem 6 asserts in effect the Hardy-Littlewood inequalities for the maximal function. Anon, we shall see that Theorem 6 (essentially) implies the maximal theorem for semigroups, in a rather general case, which we have proved with much machinery and effort.

To prove Theorem 6 we need the following:

LEMMA 1: *Let* $f_1, f_2, \ldots, f_n$ *be a finite sequence of functions on* (M, dx), *where* f_j *is measurable with respect to* $\mathcal{F}_j$. *Suppose that* $f_j = E(f_n|\mathcal{F}_j)$. *Then* $f_n^*(x) \equiv \sup_{j\leq n} f_j(x)$ *satisfies the inequalities*

(a)′ $m\{x| f_n^*(x) > a\} \leq \frac{A}{a}\|f_n\|_1$, *with* A *independent of* n .

(b)′ $\|f_n^*\|_p \leq A_p\|f_n\|_p$, *with* A *independent of* n.

Proof: We may as well suppose that $f_n \geq 0$, so that all f_j are positive. Let $S_a = \{x \in M| f_n^*(x) > a\} = \{x \in M|$ some $f_j(x)$ exceeds $a\}$. Then

$S_\alpha = \cup_{i=1}^n S_\alpha^{(j)}$ where $S_\alpha^{(j)} \equiv \{x \in M \mid f_j(x) > \alpha,$ but $f_\ell(x) \leq \alpha$ for $\ell < j\}$, and the sets $S_\alpha^{(j)}$ are pairwise disjoint.

The crucial point is that $S_\alpha^{(j)}$ is measurable with respect to the field $\mathcal{F}_j$. From this fact we deduce that

$$\int_{S_\alpha} f_n(x)dx = \sum_{j=1}^n \int_{S_\alpha^{(j)}} f_n(x)dx = \sum_{j=1}^n \int_{S_\alpha^{(j)}} E(f_n \mid \mathcal{F}_j)(x)dx$$

$$= \sum_{j=1}^n \int_{S_\alpha^{(j)}} f_j(x)dx \geq \sum_{j=1}^n \alpha\, m(S_\alpha^{(j)}) = \alpha\, m(S_\alpha) .$$

In other words $m\{x \in M \mid f_n^*(x) > \alpha\} \leq \frac{1}{\alpha} \int_{\{x \in M \mid f_n^*(x) > \alpha\}} f_n(x)dx$, which prove (a).′ Assertion (b)′ is a trivial consequence of (a)′ and the following special case of the Marcinkiewicz interpolation theorem.

LEMMA 2: *Let* T *be a mapping from* $L_p(M, dx)$ *to* $L_p(M, dx)$ *for all* p $(1 \leq p \leq +\infty)$. *Suppose that* T *is a sub-linear, i.e.,*

$$|T(f_1 + f_2)(x)| \leq |Tf_1(x)| + |Tf_2(x)|$$

for all f_1, f_2, *and all* $x \in M$. *Furthermore, suppose that* T *satisfies the inequalities*

(1) $m\{x \in M \mid |Tf(x)| > \alpha\} \leq \frac{A}{\alpha} \|f\|_1$ (i.e., T has *weak-type* (1, 1)).

(2) $\|Tf\|_\infty \leq A\|f\|_\infty$.

Then for every $f \in L_p$ $(1 < p < +\infty)$, Tf *belongs to* L_p, *and the inequality*

(3) $\|Tf\|_p \leq A_p \|f\|_p$

is valid, where A_p *depends only on* A *and* p.

Proof: For convenience, suppose $A = \frac{1}{2}$, and denote the measure of a set E by $|E|$.

Let F be any function in $L_p(M, dx)$ $(1 \leq p < +\infty)$. A moment's thought shows that $\|F\|_p^p = -\int_0^\infty \alpha^p d|\{x \in M \mid |F(x)| > \alpha\}|$. Integration by parts yields

$$\|F\|_p^p = (p)\int_0^\infty \alpha^{p-1}|\{x \in M| \ |F(x)| > \alpha\}|\,d\alpha \ .$$

In particular, picking an $f \geq 0$, we have

(4)
$$\|Tf\|_p^p = p\int_0^\infty \alpha^{p-1}|\{m \in M| \ |Tf(x)| > \alpha\}|\,d\alpha \ .$$

To estimate $|\{x \in M| \ |Tf(x)| > \alpha\}|$, write $f = g_\alpha + h_\alpha$ where

$$g_\alpha(x) = \begin{cases} f(x) & \text{if } f(x) < \alpha \\ \alpha & \text{if } f(x) \geq \alpha \end{cases} \quad \text{and} \quad h_\alpha(x) = \begin{cases} 0 & \text{if } f(x) < \alpha \\ f(x)-\alpha & \text{if } f(x) \geq \alpha \ . \end{cases}$$

Clearly $\|g_\alpha\|_\infty \leq \alpha$ and $|\{x \in M| \ h_\alpha(x) > \lambda\}| = |\{x \in M| \ f(x) > \lambda + \alpha\}|$.
Now, we have

$$|\{x \in M| \ |Tf(x)| > \alpha\}| \leq |\{x \in M| \ |Tg_\alpha(x)| > \tfrac{\alpha}{2}\}| + |\{x \in M| \ |Th_\alpha(x) > \tfrac{\alpha}{2}\}|$$

by the sublinearity of T. The first term on the right-hand side of the inequality is zero, by virtue of (2). The second term is at most $1/\alpha \int_M h_\alpha(x)dx$ (by (1)), which is equal to $1/\alpha \int_0^\infty |\{x| \ h_\alpha(x) > \lambda\}|\,d\lambda$ $= 1/\alpha \int_\alpha^\infty |\{x| \ f(x) > \lambda\}|\,d\lambda$. Thus

$$|\{x \in M| \ |Tf(x)| > \alpha\}| \leq \frac{1}{\alpha}\int_\alpha^\infty |\{x| \ f(x) > \lambda\}|\,d\lambda \ .$$

Applying this to equation (4), we obtain

$$\|Tf\|_p^p \leq p\int_0^\infty \alpha^{p-2}\int_\alpha^\infty |\{x \in M| \ f(x) > \lambda\}|\,d\lambda \, d\alpha$$

$$= p\int_0^\infty \int_0^\lambda \alpha^{p-2}\,d\alpha \ |\{x \in M| \ f(x) > \lambda\}|\,d\lambda \ \text{(by Fubini's}$$
theorem)

$$= \frac{p}{p-1}\int_0^\infty \lambda^{p-1}|\{x \in M| \ f(x) > \lambda\}|\,d\lambda = (\frac{p}{p-1})\|f\|_p^p \ .$$

This completes the proof of (3).

Proof of Theorem 6: By Lemma 1, $\|\sup_{j\leq n} f_j(\cdot)\|_p \leq A\|f\|_p$ $(1 < p \leq +\infty)$, where A is independent of n. Since the sequence $\{\sup_{j\leq n} f_j(\cdot)\}$ increases to $f^*(\cdot)$, Theorem 6, part (b), follows from the monotone convergence theorem.

Similarly, we can easily deduce Theorem 6, part (a), from (a)′ of Lemma 1. QED.

REMARKS: 1. From Theorem 6 we can prove, by the usual arguments, that for $f \in L_p$ $(1 \leq p \leq +\infty)$ $\{f_j\}$ converges almost everywhere. Details are left to the reader.

2. Lemma 1 also implies a result, analogous to Theorem 6, on "reverse martingales": Suppose we are given an infinite descending sequence of sigma-fields, $\cdots \subseteq \mathcal{F}_{n+1} \subseteq \mathcal{F}_n \subseteq \cdots \subseteq \mathcal{F}_2 \subseteq \mathcal{F}_1$. For $f \in L_1(M, \mathfrak{M}, dx)$, set $E_n(f) = f_n = E(f|\mathcal{F}_n)$. Then $f^*(x) \equiv \sup_n |f_n(x)|$ satisfies inequalities (a) and (b) of Theorem 6.

The proof of this result is exactly the same as the proof of Theorem 6.

Later, we shall see that the maximal theorem for martingales, Theorem 6, implies a general case of the maximal theorem for symmetric diffusion semigroups. Furthermore, it will be shown that the Littlewood-Paley inequalities for semigroups can be deduced in part from appropriate results on martingales.

Section 2. The inequalities for martingales.

We turn to the problem of formulating and proving a martingale version of the Littlewood-Paley inequalities.

As before, we are given an increasing family of sigma-fields $\mathcal{F}_1 \subseteq \mathcal{F}_2 \subseteq \cdots$ from which we form the operators $E_n : f \to E(f|\mathcal{F}_n)$. For convenience, set $E_0 = 0$.

THEOREM 7: *Suppose that* $a = (a_1, a_2, \ldots)$ *is any sequence of numbers such that* $|a_j| \leq 1$ *for all* j. *Set*

$$T_a(f) = \sum_{k=1}^{\infty} a_k(E_k - E_{k-1})f \qquad \textit{for} \quad f \in L_2 .$$

T_a *is well-defined for* $f \in L_2$, *by virtue of the orthogonality of the* $(E_k - E_{k-1})f$. *Then*

1. $\|T_a f\|_p \leq A_p \|f\|_p$ *for* $f \in L_2 \cap L_p$ $(1 < p < +\infty)$.
2. T_a *has weak-type* (1, 1), *i.e.,* $m\{x \mid T_a f(x) > \lambda\} \leq \frac{A}{\lambda}\|f\|_1$ *for* $f \in L_1 \cap L_2$.
3. *If we define the "Littlewood-Paley" function* $G(f)(x) \equiv (\Sigma_{k=1}^{\infty} |E_k f(x) - E_{k-1} f(x)|^2)^{1/2}$, *then for* $f \in L_2 \cap L_p$,
$$\|G(f)\|_p \leq A_p \|f\|_p, \qquad (1 < p < +\infty).$$
4. *Let* $E_\infty f = \lim_{n\to\infty} E_n f$. *Then* $\|E_\infty f\|_p \leq A_p \|G(f)\|_p$.

The constants A *and* A_p *appearing in* 1. *and* 2. *do not depend on the particular sequence* (a).

REMARKS: Theorem 7 was first proved for the dyadic interval case, by Paley, who formulated it as a theorem on Walsh-Paley series in his 1931 paper [17]. The involved proof he gave can be carried over to the general case. However, we shall present here the proof of Gundy [28], which is modern in viewpoint, and is based on a lemma analogous to the Calderon-Zygmund decomposition. The proof of the general theorem 7 is given in Burkholder [24]; see also Austin [23].

Gundy's Lemma is the following:

Let $f \geq 0$ *belong to* L_1. *Given* $\lambda > 0$, *we can write* $f = g + h + k$, *where*

(a) $m\{x \mid \sup_n |E_n(g)(x)| > 0\} \leq \frac{K}{\lambda}\|f\|_1$, *and* $\|g\|_1 \leq K\|f\|_1$.

(b) $\|\Sigma_{n=1}^{\infty} |E_n(h) - E_{n-1}(h)\|_1 \leq K\|f\|_1$. *In particular* $\|h\|_1 \leq K\|f\|_1$.

(c) $\|k\|_\infty \leq K\lambda$ *and* $\|k\|_1 \leq K\|f\|_1$.

From Gundy's lemma, we can easily prove parts 1. and 2. of Theorem 7. For if $f \in L_1 \cap L_2$, and $f = g + h + k$ as in the lemma, then we have

$$m\{x \mid T_a f(x) > \lambda\} \leq m\{x \mid T_a g(x) > \tfrac{\lambda}{3}\} + m\{x \mid T_a h(x) > \tfrac{\lambda}{3}\}$$
$$+ m\{x \mid T_a k(x) > \tfrac{\lambda}{3}\} \equiv \text{I.} + \text{II.} + \text{III.}$$

Now I. $\leq m\{x \mid \sup_n |E_n(g)(x) \neq 0\} \leq \frac{K}{\lambda}\|f\|_1$. To estimate II, we note that $|T_a(h)(x)| \leq \Sigma_n |(E_n - E_{n-1})h(x)|$. So

$$\text{II.} \leq m\{x \mid \Sigma_n |(E_n - E_{n-1})f(x)| > \tfrac{\lambda}{3}\} \leq \frac{K}{\lambda}\|f\|_1$$

by (b) of Gundy's lemma. Finally, III. $\leq \frac{K}{\lambda}\|f\|_1$. For by (c) of the lemma, we have the inequality $\|k\|_2^2 \leq K\lambda\|f\|_1$. Since T_a is a bounded operator on L_2,

$$m\{x \mid T_a(k)(x) \geq \tfrac{\lambda}{3}\} \leq \frac{9}{\lambda^2}\|T_a k\|_2^2 \leq \frac{9}{\lambda^2}\|k\|_2^2 \leq \frac{K}{\lambda}\|f\|_1$$

Putting together our estimates for I., II., and III., we find that

$$m\{x \mid T_a(f) > \lambda\} \leq \frac{K}{\lambda}\|f\|_1$$

which is exactly part 2 of Theorem 7.

The operator T_a thus has weak-type (1, 1) and strong type (2, 2) (i.e., T_a is a bounded operator on L_2). By the Marcinkiewicz interpolation theorem, T_a is a bounded linear operator on L_p $(1 < p \leq 2)$. (The basic idea of the proof of the Marcinkiewicz interpolation theorem is already indicated in Lemma 2 in the proof of Theorem 6 above. A detailed discussion of the Marcinkiewicz interpolation theorem may be found in Zygmund's *Trigonometric Series*, Chapter XII, [20].) That T is bounded on L_p $(2 \leq p < +\infty)$ follows from the usual duality argument.

Proof of Gundy's Lemma: We are given an L_1 function f (say, $f \geq 0$) and a $\lambda > 0$. Let $f_n = E_n(f) = E(f \mid \mathcal{F}_n)$. The decomposition of f will be carried out using the $\{f_n\}$, and the notion of a *stopping time*, which we now define.

Suppose $r(x)$ is a positive integer-valued function on the measure space $(M, \mathfrak{M}, dx)$ such that $\{x \mid r(x) = n\}$ is measurable not only with respect to $\mathfrak{M}$, but with respect to $\mathcal{F}_n$, for each $n \geq 1$. $r(\cdot)$ is then called a stopping time.

If $r(x)$ is a stopping time, then

$$(*) \qquad \int_M f(x)dx = \int_M f_{r(x)}(x)dx .$$

For the left-hand side of (*) is just

$$\sum_{j=1}^{\infty} \int_{\{x \mid r(x)=j\}} f_j(x)dx \equiv \sum_{j=1}^{\infty} \int_{\{x \mid r(x)=j\}} E(f \mid \mathcal{F}_j)dx$$

$$= \sum_{j=1}^{\infty} \int_{\{x \mid r(x)=j\}} f(x)dx = \int_M f(x)dx .$$

Equation (*) generalizes the identity $\int_M f_n(x)dx = \int_M f(x)dx$.

We can construct a new martingale from $\{f_n\}$ and $r(x)$. Simply set $f'_n(x) = f_{\min(n,r(x))}(x)$. $\{f'_n\}$ is called the *stopped martingale* defined by $\{f_n\}$ and r. (For each x, $\{f'_n(x)\}$ looks just like $\{f_n(x)\}$ until time $r(x)$, when $\{f'_n\}$ "stops".) The proof that $\{f'_n\}$ form a martingale, i.e., $f'_n = E_n(f'_{n+1})$ resembles the proof of (*), and is left to the reader.

As a slight extension of the definition of stopping time, we allow $r(x)$ to take on the value $+\infty$. Equation (*) still holds if we define $f_\infty(x) = f(x)$, and $\{f'_n\}$ is still a martingale. (In fact $f'_n = E_n(f_{r(x)}(x))$.)

Recall that we are trying to prove Gundy's Lemma. The easiest of the three parts of the decomposition is g, which will be defined by $g(x) = f(x) - f_{t(x)}(x)$ where $t(x)$ is a particular stopping time. To define t, we let $r(x) = \inf\{n \mid f_n(x) > \lambda\}$. $r(x)$ is a stopping time, since

$$\{x \mid r(x) = n\} = \{x \mid f_1(x), \ldots, f_{n-1}(x) \leq \lambda \text{ but } f_n(x) > \lambda\} \in \mathcal{F}_n .$$

Next write $f_n(x) = \sum_{k=1}^{n} \phi_k(x)$ where $\phi_k = f_k - f_{k-1}$ and $f_0 \equiv 0$;

and set $\varepsilon_n(x) = \phi_n(x)\,\chi_{\{y|r(y)=n\}}(x)$. Obviously $\varepsilon_n \geq 0$. (Think about it for a moment.) Define a new stopping time s by $s(x) = \inf\{n \mid \sum_{k=0}^{n} E(\varepsilon_{k+1} \mid \mathcal{F}_k)(x) > \lambda\}$. (The reader may check that s is, in fact, a stopping time.) Now set $t(x) = \min(r(x), s(x))$. t is a stopping time, since it is the minimum of two stopping times.

I claim that $m\{x \mid t(x) < +\infty\} \leq \frac{K}{\lambda}\,\|f\|_1$, where K is a universal constant. First of all, $\{x \mid r(x) \neq +\infty\} = \{x \mid \sup_n f_n(x) > \lambda\}$, so that $m\{x \mid r(x) < +\infty\} \leq \frac{K}{\lambda}\,\|f\|_1$ by the martingale maximal theorem. Similarly, $\{x \mid \sum_{k=0}^{\infty} E(\varepsilon_{k+1} \mid \mathcal{F}_k)(x) > \lambda\}$ and

$$\int_M \sum_{k=0}^{\infty} E(\varepsilon_{k+1} \mid \mathcal{F}_k)(x)\,dx = \int_M \sum_{k=0}^{\infty} \varepsilon_{k+1}(x)dx$$

$$= \sum_{k=0}^{\infty} \int_{\{x|r(x)=k+1\}} (f_{k+1}(x) - f_k(x))dx$$

$$\leq \sum_{k=0}^{\infty} \int_{\{x|r(x)=k+1\}} f_{k+1}(x)dx \quad (\text{since } f \geq 0)$$

$$= \int_{\{x|r(x)<+\infty\}} f_{r(x)}(x)dx \leq \int_M f_{r(x)}(x)dx = \int_M f(x)dx$$

(by (*)) $$= \|f\|_1\,.$$

Hence $m\{x \mid s(x) < +\infty\} \leq \frac{K}{\lambda}\,\|f\|_1$ with $K = 1$. Finally,

$$m\{x \mid t(x) < +\infty\} \leq m\{x \mid r(x) < +\infty\} + m\{x \mid s(x) < +\infty\}$$

$$\leq \frac{K'}{\lambda}\,\|f\|_1 + \frac{K''}{\lambda}\,\|f\|_1 \leq \frac{K}{\lambda}\,\|f\|_1\,.$$

As advertised, define $g(x) = f(x) - f_{t(x)}(x)$. Immediately, we see that $\{x \mid g(x) \neq 0\} \subseteq \{x \mid t(x) \neq +\infty\}$ so that $m\{x \mid g(x) \neq 0\} \leq \frac{K}{\lambda}\,\|f\|_1$. With only a little more work we can prove part a of the splitting lemma. For,

$E_n(g) = f_n - \tilde{f}_n$, where $\{\tilde{f}_n\}$ is the stopped martingale defined by $\{f_n\}$ and the stopping time $t(x)$. So $\{x \mid \sup_n |E_n(g)(x)| \neq 0\} \subseteq \{x \mid t(x) \neq +\infty\}$, which shows that $m\{x \mid \sup_n |E_n(g)(x) \neq 0\} \leq \frac{K}{\lambda} \|f\|_1$. Part c is proved.

Note that $\|g\|_1 \leq \|f\|_1 + \|f_{t(x)}(x)\|_1 \leq 2\|f\|_1$.

Since we have to wind up with $g + h + k = f$, we are left with the job of constructing functions h and k such that $h(x) + k(x) = f_{t(x)}(x)$ and conditions b and c hold. Shall we have a go at it?

Begin with the formula $\tilde{f}_n = \Sigma_{j=1}^n \phi_j \chi_{\{y \mid t(y) \geq j\}}$, which follows from decipherment of notation. Now $\{y \mid t(y) \geq j\} = \{y \mid r(y) \geq j\} \cap \{y \mid s(y) \geq j\}$. Since $\phi_j \chi_{\{y \mid r(y) = j\}} = \varepsilon_j$, we can write

$$\tilde{f}_n = \sum_{j=1}^{n} (\gamma_j + \varepsilon_j)\chi_{\{y \mid s(y) \geq j\}} \text{ where } \gamma_j \equiv \phi_j \chi_{\{y \mid r(y) > j\}} .$$

Set $h_n(x) = \Sigma_{j=1}^n (\varepsilon_j - E(\varepsilon_j \mid \mathcal{F}_{j-1})) \cdot \chi_{\{y \mid s(y) \geq j\}} \equiv \Sigma_{j=1}^n \psi_j$ and

$k_n(x) = \Sigma_{j=1}^n (\gamma_j + E(\varepsilon_j \mid \mathcal{F}_{j-1})) \cdot \chi_{\{j \mid s(y) \geq j\}}$. Obviously, $h_n + k_n = \tilde{f}_n$, so that in the distant future, when we learn that $\{h_n\}$ and $\{k_n\}$ converge in the L_1-norm, the limits h and k will satisfy $h(x) + k(x) = f_{t(x)}(x)$, i.e., $g + h + k = f$.

I claim that $\{h_n\}$ and $\{k_n\}$ are martingales, i.e., $E_n(h_{n+1}) = h_n$ and $E_n(k_{n+1}) = k_n$. Since $k_n = \tilde{f}_n - h_n$, we need only check the martingaledom of $\{h_n\}$. But $E_n(h_{n+1}) = E_n(\Sigma_{j=1}^{n+1} \psi_j) = E_n(\psi_{n+1}) + \Sigma_{j=1}^n \psi_j$ (since by inspection, ψ_j is $\mathcal{F}_j$-measurable) $= E_n(\psi_{n+1}) + h_n$. We are thus reduced to showing that $E_n(\psi_{n+1}) = 0$.

By definition $E_n(\psi_{n+1}) = E_n((\varepsilon_{n+1} - E_n(\varepsilon_{n+1})) \cdot \chi_{\{y \mid s(y) \geq n+1\}})$

$= (E_n(\varepsilon_{n+1} - E_n(\varepsilon_{n+1}))) \cdot \chi_{\{y \mid s(y) \geq n+1\}}$ (since $\{y \mid s(y) \geq n+1\}$

$= M - \cup_{j=1}^n \{y \mid s(y) = j\}$ is measurable with respect to $\mathcal{F}_n$) $= (E_n(\varepsilon_{n+1}) -$

$E_n((\varepsilon_{n+1})) \cdot \chi_{\{y \mid s(y) \geq n+1\}} = 0$. This proves that $\{h_n\}$ and $\{k_n\}$ are martingales.

Next, we shall verify that

$$(**)\qquad \sum_{j=1}^{\infty} \|\psi_j\|_1 \leq K\|f\|_1 .$$

This will show that $\{h_n\}$ converges in L_1 to a limit h. $E_n h = h_n$ since $\{h_n\}$ is a martingale, so $E_n h - E_{n-1} h = \psi_n$, which implies that $\Sigma_{j=1}^{\infty} \|E_j h - E_{j-1} h\|_1 \leq K\|f\|_1$. The construction of h and the proof of part b of the lemma are therefore consequences of (**).

Written out in full, (**) says that

$$\sum_j \int_M |\varepsilon_j - E(\varepsilon_j|\mathcal{F}_{j-1})| \cdot \chi_{\{y|s(y)\geq j\}} dx \leq K\|f\|_1 .$$

But since $\varepsilon_j \geq 0$, the left-hand side of this inequality is dominated by

$$\sum_j \int_M \varepsilon_j \chi_{\{y|s(y)\geq j\}} dx + \sum_j \int_M E(\varepsilon_j|\mathcal{F}_{j-1}) \cdot \chi_{\{y|s(y)\geq j\}} dx$$

$$= 2\sum_j \int_M \varepsilon_j(x)\chi_{\{y|s(y)\geq j\}} dx \text{ (by definition of conditional expectation)}$$

$$= 2\int_M \sum_j \varepsilon_j(x)dx =$$

$$= 2\int_M \sum_{j=1}^{\infty} (f_j(x) - f_{j-1}(x)) \cdot \chi_{\{y|r(y)=j\}}(x)dx \text{ (by definition of } \varepsilon_j)$$

$$\leq 2\int_M \sum_{j=1}^{\infty} f_j(x)\chi_{\{y|r(y)=j\}}(x)dx = 2\int_{\{y|r(y)<+\infty\}} f_{r(x)}(x)dx$$

$$\leq 2\int_M f_{r(x)}(x)dx = 2\|f\|_1 \text{ by } (*) .$$

Therefore (**) is verified with $K = 2$.

So far we haven't used any information on s(x) except that it is a stopping time. We now finish off the proof of the lemma by exploiting the properties of s(x) to prove c.

By the properties of g and h already demonstrated, $k = f - g - h$ has L_1 norm $\|k\|_1 \leq K\|f\|_1$; and we have the representation

$$k = \sum_{j=1}^{\infty} (\gamma_j + E(\varepsilon_j|\mathcal{F}_{j-1})) - \chi_{\{y|s(y)\geq j\}},$$

valid pointwise almost everywhere. Part 3 of the lemma says that

$$|k(x)| \leq K\lambda$$

almost everywhere, so it is surely enough to prove that

$$(\alpha) \qquad \Big\|\sum_{j=1}^{\infty} \gamma_j - \chi_{\{y|s(y)\geq j\}}\Big\|_\infty \leq K\lambda$$

and

$$(\beta) \qquad \Big\|\sum_{j=1}^{\infty} E(\varepsilon_j|\mathcal{F}_{j-1})\chi_{\{y|s(y)\geq j\}}\Big\|_\infty \leq K\lambda .$$

(α) follows from the computation

$$\sum_{j=1}^{\infty} \gamma_j(x)\,\chi_{\{y|S(y)\geq j\}} = \sum_{j=1}^{\infty} \phi_j(x)\chi_{\{y|r(y)>j\}}(x)\cdot\chi_{\{y|s(y)\geq j\}}(x)$$

$$= \sum_{j=1}^{\min(r(x)-1,\,s(x))} \phi_j(x) = f_{\min(r(x)-1,\,s(x))}(x)$$

which has absolute value at most λ, by definition of r(x).

Similarly, (β) follows from the computation

$$0 \leq \sum_{j=1}^{\infty} E(\varepsilon_j|\mathcal{F}_{j-1})(x)\cdot\chi_{\{y|s(y)\geq j\}}(x)$$

$$= \sum_{j=1}^{s(x)} E(\varepsilon_j|\mathcal{F}_{j-1})(x) = \sum_{\ell=0}^{s(x)-1} E(\varepsilon_{\ell+1}|\mathcal{F}_\ell)(x) \leq \lambda,$$

by definition of s(x). This completes the proof of Gundy's lemma. QED.

Recall that we are trying to prove Theorem 7, and that we already established parts 1 and 2 of the theorem, by using the splitting lemma. To carry on, we need another of the basic tools of Fourier analysis—the family $\{r_k\}$ of *Rademacher functions.* If k is a non-negative integer, then r_k is the function on $[0,1]$ defined by

$$r_k(t) = \begin{cases} 1 & \text{if } j/2^k \le t < (j+1)/2^k, \quad j \text{ even} \\ -1 & \text{if } j/2^k \le t < (j+1)/2^k, \quad j \text{ odd.} \end{cases}$$

The $\{r_k\}$ form an orthonormal system on $[0,1]$, which is, however, very far from being complete.

Suppose that $F(t) = \Sigma_{k=0}^{\infty} a_k r_k(t)$, where $\Sigma_{k=0}^{\infty} |a_k|^2 < +\infty$. Then of course $F \in L_2$ and $\|F\|_2 = (\Sigma_{k=0}^{\infty} |a_k|^2)^{1/2}$. But we can say much more. For any p $(1 \le p < +\infty)$, $F \in L_p[0,1]$, and

$$(*) \qquad B_p\left(\sum_{k=0}^{\infty} |a_k|^2\right)^{1/2} \le \|F\|_p \le A_p\left(\sum_{k=0}^{\infty} |a_k|^2\right)^{1/2}$$

where the constants A_p and B_p depend only on p.

For a proof of (*), see Zygmund's *Trigonometric Series*, Vol. I. Chapter V, [20].

We can now prove part 3. of the theorem, by using part 1. and inequality (*) for the Rademacher functions. Part 1 says that $\int_M |T_a f(x)|^p dx \le A_p \|f\|_p^p$ where $T_a f = \Sigma_{k=1}^{\infty} a_k(E_k f - E_{k-1} f)$ and $a = (a_k)$ is any sequence of norm 1. For $t \in [0,1]$, let $a_k = r_k(t)$. We obtain the inequality

$$\int_M \left|\sum_{k=1}^{\infty} (E_k f(x) - E_{k-1} f(x)) \cdot r_k(t)\right|^p dx \le A_p \|f\|_p^p$$

with A_p independent of t. Integrating in t, and changing the order of integration, yields

$$\int_M \left[\int_0^1 \left|\sum_{k=1}^{\infty} (E_k f(x) - E_{k-1} f(x)) \cdot r_k(t)\right|^p dt\right] dx \le A_p \|f\|_p^p .$$

By inequality (*),the expression in brackets is approximately

$$A_p' \cdot \left(\sum_{k=1}^{\infty} |E_k f(x) - E_{k-1} f(x)|^2 \right)^{p/2} \equiv A_p' |G(f)(x)|^p .$$

Therefore, $\int_M |G(f)(x)|^p dx \leq A_p \|f\|_p^p$ for all $f \in L_p(M, dx)$. This completes the proof of part 3 of Theorem 7.

Part 4 comes from part 3 by the usual duality argument, based on the fact that G is an isometry on L_2. Thus all parts of Theorem 7 are proved. Q.E.D.

Section 3. An additional "max" inequality.

We have proved two big theorems on martingales—the "Paley inequality" and the maximal theorem. There remains one more result, and after we get it out of the way, we can come (finally!) to the link between semigroups and martingales, that will enable us to prove the general semigroup form of the Littlewood-Paley inequality.

THEOREM 8. *Given* $\mathcal{F}_1 \subseteq \mathcal{F}_2 \subseteq \cdots$ *as before, let* E_k *denote the conditional expectation operator with respect to* $\mathcal{F}_k$. *Suppose that* $\{f_k\}$ *is any sequence of functions on* (M, dx), *where* f_k *is not assumed to be* $\mathcal{F}_k$-*measurable; and let* $\{n_k\}$ *be any sequence of positive integers. Then*

$$\left\| \left(\sum_k |E_{n_k}(f_k)|^2 \right)^{1/2} \right\|_p \leq A_p \left\| \left(\sum_k |f_k|^2 \right)^{1/2} \right\|_p \qquad (1 < p < +\infty)$$

where A_p *depends only on* p.

Proof: The theorem has an easy proof. Let $L_p(\ell_q)$ denote the Banach space of all sequences of functions, $\{f_k\}$, for which the norm

$$\|(f_k)\| \equiv \left(\int_M \left(\sum_k |f_k(x)|^q \right)^{p/q} dx \right)^{1/p}$$

is finite. (If $q = +\infty$ we make the obvious modification

$$\|(f_k)\| \equiv \left(\int_M \left(\sup_k |f_k(x)| \right)^p dx \right)^{1/p}, \;)$$

$L_p(\ell_q)$ is really very much like L_p. For example, the dual space of $L_p(\ell_q)$ is $L_{p'}(\ell_{q'})$, under the pairing $<(f_k),(g_k)> = \int_M \Sigma_k f_k(x) g_k(x)dx$, where $1/p' + 1/p = 1/q' + 1/q = 1$, provided that $p \neq +\infty$, $q \neq +\infty$.

We shall use the following generalization of the Riesz convexity theorem: Let T be a linear operator which maps sequences of functions to sequences of functions. Suppose that T is bounded as an operator from $L_{p_0}(\ell_{q_0})$ to itself, and as an operator from $L_{p_1}(\ell_{q_1})$ to itself. Then T is also bounded as an operator from $L_{p_t}(\ell_{q_t})$ to itself, where

$$\frac{1}{p} = \frac{(1-t)}{p_0} + \frac{t}{p_1}$$

and

$$\frac{1}{q_t} = \frac{(1-t)}{q_0} + \frac{t}{q_1} \cdot \qquad (0 \leq t \leq 1)\ .$$

The proof of this theorem is very similar to that of the Riesz convexity theorem. A full proof is found in a paper of Benedeck-Panzone, *The Spaces L^p with Mixed Norm* (Duke Math. J. 1961, p. 301--324), [21]. See also Calderón [39].

Now, consider the operator T, which sends the sequence $\{f_k\}$ of functions, to the sequence $\{E_{n_k} f_k\}$. T is a bounded operator on $L_p(\ell_p)$, since

$$\int_M \left(\sum_k |E_{n_k} f_k(x)|^p \right)^{p/p} dx = \sum_k \int_M |E_{n_k} f_k(x)|^p dx \leq \sum_k \int_M |f_k(x)|^p dx$$
$$= \int_M \left(\sum_k |f_k(x)|^p \right)^{p/p} dx\ .$$

On the other hand, T is a bounded operator on $L_p(\ell_\infty)$ if $1 < p \leq +\infty$. This is because

$$\int_M |\sup_k E_{n_k} f_k(x)|^p dx \leq \int_M |\sup_{n,k} E_n f_k(x)|^p dx \leq \int_M |\phi^*(x)|^p dx$$
$$= A_p \int (\sup_k |f_k(x)|)^p dx\ ,$$

(where * denotes the maximal function, and $\phi(x) \equiv \sup_k |f_k(x)|$), by the maximal theorem.

We can now apply the generalized Riesz convexity theorem to conclude that T is bounded on $L_p(\ell_q)$ if $1 < p \leq q \leq +\infty$. In particular, if $1 < p \leq 2$, then T is bounded on $L_p(\ell_2)$, which is precisely the statement of the theorem!

The case $2 \leq p < +\infty$ follows by an obvious duality argument involving $L_p(\ell_q)$-spaces. Q.E.D.

REMARKS. The result of Theorem 8 does not hold when either $p = 1$ or $p = \infty$, and in fact the true order of growth of the bound A_p is $O(p^{1/2})$ or $p \to \infty$, and $O((p-1)^{-1/2})$, as $p \to 1$. This indicates that the theorem cannot be entirely trivial.

The fact $A_p \leq Ap^{1/2}$ as $p \to \infty$ follows by an examination of the bounds arising from the interpolation argument. To show that in fact $A_p > Ap^{1/2}$ in general, let $E_1, E_2, \ldots, E_k \ldots$ arise from the "dyadic interval" expectations and set $f_k(x) = 1$ if $2^{-k-1} < x \leq 2^{-k}$, $f_k(x) = 0$) otherwise. Then $(\Sigma |f_k(x)|^2)^{1/2} \equiv 1$, while

$$(\Sigma\, |E_k(f_k)|^2)^{1/2} \geq c(\log \frac{1}{x})^{1/2}\,, \qquad 0 \leq x \leq 1\,.$$

COROLLARY: *If* $\{f_k\}$ *is any sequence of functions on* (M, dx), *then*

$$\Big\| \Big(\sum_k |f_k^*|^2 \Big)^{1/2} \Big\|_p \leq A_p \Big\| \Big(\sum_k |f_k|^2 \Big)^{1/2} \Big\|_p \qquad (1 < p \leq 2).$$

Proof: Carry out the proof of theorems, using the operator $T(\{f_k\})(x) = \{E_{n(k,x)} f_k(x)\}$, where $n(\cdot\,,\cdot)$ is an arbitrary (measurable) positive-integer-valued function of (n, k). The corollary now follows by the argument we used to carry out the interpolation in the proof of the semigroup maximal theorem. (Note that we cannot use the duality argument here, since $\{f_k\} \to \{f_k^*\}$ is non-linear, and T is non-self-adjoint.) Q.E.D.

Interesting question: Is this true for $2 < p < +\infty$?*

All our martingale inequalities are also valid for reverse martingales.

---

* See Appendix (1985).

Section 4. The Link Between Martingales and Semigroups

Since (our) martingales depend on a discrete parameter n, and semigroups depend on a continuous time parameter t, it becomes expedient for us to discretize our semigroups. So we shall study the powers of an operator Q on $L_p(M, dx)$ which satisfies the axioms

(I) $\|Qf\|_p \leq \|f\|_p$, $(1 \leq p \leq +\infty)$.

(II) Q is self-adjoint on L_2.

(III) $Qf \geq 0$ if $f \geq 0$.

(IV) $Q1 = 1$.

THEOREM 9 (Rota): *There is a huge measure space (Ω, β, P), a collection of sigma-fields $\cdots \subseteq \mathcal{F}_{n+1} \subseteq \mathcal{F}_n \subseteq \cdots \subseteq \mathcal{F}_1 \subseteq \mathcal{F}_0$, and another sigma-field $\hat{\mathcal{F}}_0$, all contained in β, with the following properties:*

(1) *The measure spaces $(M, \mathfrak{M}, dx)$ and $(\Omega, \hat{\mathcal{F}}_0, P)$ are isomorphic under a natural mapping $i : \Omega \to M$. The induced isomorphism of $L_p(\Omega, \mathcal{F}_0, P)$ with $L_p(M, \mathfrak{M}, dx)$ will also be denoted by i.*

(2) *Let $f \in L_p(\Omega, \hat{\mathcal{F}}_0, P)$: Then $Q^{2n}(i(f)) = (\hat{E} E_n f)$, where $\hat{E}$ and E_n are the conditional expectation operators for $\hat{\mathcal{F}}_0$ and $\mathcal{F}_n$ (respectively).*

Thus, the operator Q is associated with a reverse martingale, $\{E_n\}$.

From this result and the martingale maximal theorem, we can easily deduce the maximal theorem for semigroups, which satisfy (I), (II), (III), and (IV). For, let $\{T^t\}$ be a semigroup satisfying our axioms. Then the operator $Q = T^{1/2^{k+1}}$ satisfies axioms (I)–(IV) above. The martingale maximal theorem and (2) above that

$$\|\sup_n Q^{2n}(if)(\cdot)\|_p = \|i^{-1}(\sup_n Q^{2n}(if))\|_p = \|\sup_n \hat{E} E_n(f(\cdot)\|_p$$

$$\leq \|\hat{E}(\sup_n E_n f)\|_p \leq \|\sup_n E_n f\|_p$$

$$\leq A_p \|f\|_p ,$$

so that $\|\sup_n Q^{2n}f(\cdot)\|_p \leq A_p\|f\|_p$ for $f \in L_p(M, dx)$. In other words, $\|\sup_n T^{n/2^k}f(\cdot)\|_p \leq A_p\|f\|_p$, with A_p independent of k. Letting $k \to \infty$, we conclude from the monotone convergence theorem that

$$\|\sup_{t>0} T^t f(\cdot)\|_p \leq A_p\|f\|_p$$

(recall that $T^t f(x)$ is a continuous function of $t \in (0, +\infty)$).

This deduction was not necessary at this stage—after all, we already knew a proof of the semigroup maximal theorem (which didn't even require axioms (III) and (IV).)

But we can already surmise the power of the martingale theorems when combined with Theorem 9.

Proof of Theorem 9: (Ω, β, P) is actually the result of an old construction from the theory of Markov processes.

Imagine a particle located somewhere inside M, (say, at p_0) at time $t = 0$. At time $t = 1$ the particle jumps to some other point p_1 of M, according to some fixed probability distribution for p_1. Having reached p_1, the particle forgets that it was ever at p_0. So at time $t = 2$, the particle jumps to a point $p_2 \in M$, and the probability distribution of p_2 depends only on p_1, *not* on p_0. The process continues—at time $t = n+1$ the particle jumps from p_n to p_{n+1}, having completely forgotten where it was at times $0, 1, \ldots, n-1$.

There is a natural probability space (Ω, β, P) for this random process. Suffice it here to define Ω. A point $\omega \in \Omega$ should describe the complete history of the peripatetic particle. So it is reasonable to set ω equal to the infinite sequence $(p_0, p_1, p_2, \ldots)$. Thus, Ω consists of all possible sequences of points of M, i.e., $\Omega = M \times M \times M \times \cdots$.

Now let us return to the case of an operator Q satisfying (I)–(IV) above, and try to use the above probabilistic ideas to construct an (Ω, β, P). First of all, we agreed that $\Omega = M \times M \times M \times \cdots$. For the Borel field β we take the sigma-field generated by all sets of the form

$$(*) \qquad A_1 \times A_2 \times \cdots \times A_N \times M \times M \times M \times \cdots,$$

where the A_i are measurable subsets of M. Note that the sets of the form (*) (so-called *cylinder sets*) already form a Boolean algebra.

To start we are going to define the measure P. Let

$$S = A_1 \times A_2 \times \cdots \times A_N \times M \times M \times M \times \cdots .$$

Start with the function χ_{A_N} on M; then form $Q\chi_{A_N}$; then multiply by $\chi_{A_{N-1}}$, to obtain $\chi_{A_{N-1}} \cdot Q(\chi_{A_N})$; apply Q again, to obtain $Q(\chi_{A_{N-1}} \cdot Q(\chi_{A_*}))$; multiply this by $\chi_{A_{N-2}}$ to obtain $\chi_{A_{N-2}} \cdot Q(\chi_{A_{N-1}} \cdot Q(\chi_{A_N}))$; apply Q again. Continuing this process, we finally come to the function $\chi_{A_0} \cdot Q(\chi_{A_1} \cdot Q(\cdots(\chi_{A_{N-1}} \cdot Q(\chi_{A_N}))\cdots)$.

Set

$$P(S) = \int_M \chi_{A_0} \cdot Q(\chi_{A_1} \cdot Q(\cdots(\chi_{A_{N-1}} \cdot Q(\chi_{A_N}))\cdots)dx .$$

The reader may check that P is well-defined, non-negative, and finitely additive on the cylinder sets ((IV) is required to show that P is well-defined, since $(A_1 \times A_2 \times \cdots \times A_N) \times M \times M \times M \times \cdots$ and $(A_1 \times A_2 \times \cdots \times A_N \times M) \times M \times M \times M \times M \times \cdots$ are the same cylinder set).

It can be shown that P extends to a countably additive measure on β. The proof is rather technical, so we omit it. The demanding reader may look in the paper of Doob, *A Ratio Operator Limit Theorem* [27]. Probabilistically, this corresponds to the situation explained at the beginning of the proof, where p_0 is distributed according to the probability law $\Pr(p_0 \in A) = \int_A dx$, and where a particle at position $p_n = x$ jumps to a position $p_{n+1} \in M$ according to the probability law $\Pr(p_{n+1} \in A) = Q(\chi_A)(x)$.

Now define $\hat{\mathcal{F}}_0 = \{A_0 \times M \times M \times M \times \cdots \mid A_0$ is a measurable subset of $M\}$, and set

$$\mathcal{F}_n = \{\overbrace{M \times M \times M \times \cdots \times M}^{n+1} \times S \mid S \in \beta, \text{ so that } S \subseteq M \times M \times M \times \cdots\}.$$

Obviously $\cdots \subseteq \mathcal{F}_{n+1} \subseteq \mathcal{F}_n \subseteq \cdots \subseteq \mathcal{F}_1 \subseteq \mathcal{F}_0 = \beta$

The mapping $i: \Omega \to M$ defined by $i(x_0, x_1, x_2, \ldots) = x_0$ sets up an isomorphism of measure spaces between $(\Omega, \hat{\mathfrak{F}}_0, P)$ and $(M, \mathfrak{M}, dx)$. Thus, part (1) of Theorem 9 is verified.

In order to prove part (2), we make two claims:

(a). If $g: \Omega \to R$ is such that $g(\{x_0, x_1, x_2, \ldots\})$, depends only on x_n then $\hat{E}(g)(\{x_0, x_1, \ldots\}) = Q^n g(x_0)$.

(b). If $g: \Omega \to R$ is such that $g(\{x_0, x_1, x_2, \ldots\})$ depends only on x_0 (i.e., g is $\hat{\mathfrak{F}}_0$-measurable), then $E_n(g)(\{x_0, x_1, \ldots\}) = Q^n g(x_n)$.

From these two claims, it is obvious that $\hat{E} E_n(i^{-1}f) = i^{-1}(Q^{2n}f)$, which proves (2) of Theorem 9. Therefore, the proof of Theorem 9 is reduced to the task of checking (a) and (b).

Proof of (a): We are given a candidate, $Q^n g(x_0)$. for the conditional expectation of g with respect to $\hat{\mathfrak{F}}_0$. Since $Q^n g(x_0)$ is obviously $\hat{\mathfrak{F}}_0$-measurable, it is enough to check that

$$(*) \qquad \int_S g(x_n)\,dP(x) = \int_S Q^n g(x_0)\,dP(x)$$

for $S \in \hat{\mathfrak{F}}_0$, i.e., $S = A_0 \times M \times M \times M \times \cdots$. Both sides of (*) are equal to $\int_{A_0} Q^n g(x)\,dx$ if g is the characteristic function of a subset of M, as follows from the definition of P. On the other hand, both sides of (*) are linear in g, and well-behaved under limit processes. So (*) is valid for all g.

Note that so far we have not used the self-adjointness of Q.

Proof of (b): As in (a), the problem reduces to showing that

$$(**) \qquad \int_S g(x_0)\,dP(x) = \int_S (Q^n g)(x_n)\,dP(x)$$

where

$$S = \underbrace{M \times M \times \cdots \times M}_{n} \times A_n \times A_{n+1} \times \cdots \times A_n \times M \times M \times M \times \cdots,$$

and g is the characteristic function of a subset of M.

The left-hand-side of (**) is equal to

$$(***)\quad \int_\Omega g(x_0)\chi_{A_n}\cdot\chi_{A_{n+1}}\cdot\chi_{A_{n+2}}\cdot\chi_{A_N}\,dP(x)$$

$$\equiv \int_M g(x)Q^n(\chi_{A_n}\cdot Q(\chi_{A_{n+1}}\cdot Q(\cdots)\cdots)dx$$

whereas the right-hand-side of (**) equals

$$(****)\quad \int_\Omega (Q^n g)(x_n)\cdot\chi_{A_n}\cdot\chi_{A_{n+1}}\cdot\cdots\cdot\chi_{A_N}\,dP(x)$$

$$= \int_M Q^n[Q^n g(x)\cdot\chi_{A_n}\cdot Q(\chi_{A_{n+1}}Q(\cdots)\cdots)]dx$$

$$= \int_M [Q^n g(x)\cdot\chi_{A_n}\cdot Q(\chi_{A_{n+1}}\cdot Q(\cdots)\cdots)]dx\ ,$$

since

$$\int_M Qh(x)dx = \int_M h(x)dx$$

for any h. Setting $\phi \equiv \chi_{A_n}\cdot Q(\chi_{A_{n+1}}\cdot Q(\cdots)\cdots)$, we obtain from (***) and (****) that the left-hand side of (**) is $\int_M g(x)Q^n\phi(x)dx$, while the right-hand side of (**) is $\int_M Q^n g(x)\cdot\phi(x)dx$. These two quantities are equal, by the self-adjointness of Q. The proof of (b) is complete. Q.E.D.

Section 5. The Littlewood-Paley Inequalities in General

Recall that if $\{T^t\}$ is a semigroup satisfying axioms (I)–(IV) and $f \in L_p(M, dx)$, then $g_1(f)(x) \equiv \left(\int_0^\infty t\,|(\partial/\partial t)\,T^t f(x)|^2 dt\right)^{1/2}$. We are at last ready to prove our main result.

THEOREM 10: *If* $f \in L_p(M, dx)$, *then* $g_1(f) \in L_p(M, dx)$, *and*

$$\|g_1(f)\|_p \leq A_p \|f\|_p \qquad (1 < p < +\infty).$$

Proof: Since we want to apply martingale theory, we shall have to deduce the desired inequality from a discrete analogue. The most natural candidate for a discrete analogue is the inequality

$$(*) \qquad \left\| \left(\sum_{n=1}^{\infty} n\,|(Q^{2n} - Q^{2n-2}) f|^2 \right)^{1/2} \right\|_p \leq A_p \|f\|_p$$

(where the operator Q satisfied (I) – (IV)) and, in fact, it is easy to prove the theorem, given (*).

Associated to $\{Q^{2n}\}$ is a martingale $\{E_n\}$ on a huge measure space $(\Omega, \mathcal{B}, P)$; in accordance with Theorem 9. To facilitate passing from $\{Q^{2n}\}$ to $\{E_n\}$ and back, let us adopt the notation that S_n stands either for the operator Q^{2n}, or for the operator E_n, depending on context. Write

$$S_n = \sum_{k=0}^{n} a_k \quad \text{where } a_k \equiv S_k - S_{k-1}\,,$$

and set

$$\sigma_n = \frac{S_0 + S_1 + \cdots + S_n}{n+1}\,,$$

the Cesaro mean.

We shall see shortly, that the most we can hope to deduce from martingale theory is

$$(**) \qquad \left\| \left(\sum_{n=1}^{\infty} n\,|(\sigma_n - \sigma_{n-1}) f|^2 \right)^{1/2} \right\|_p \leq A_p \|f\|_p \,.$$

In order to prove (*) from Theorem 9, we could try to show that

$$\left\| \left(\sum_{n=1}^{\infty} n \, |(E_n - E_{n-1}) f|^2 \right)^{1/2} \right\|_p \leq A_p \|f\|_p \ ,$$

where f is defined on Ω. Unfortunately, this inequality is false. Look at the simplest possible case, $p = 2$. Our inequality reduces to

$$\sum_{n=1}^{\infty} n \, \|(E_n - E_{n-1}) f\|_2^2 \leq A_2 \|f\|_2^2 \ .$$

But this is preposterous! For the $\{(E_n - E_{n-1}) f\}$ are orthogonal, so that the most we could ever hope for is $\Sigma_{n=1}^{\infty} \|(E_n - E_{n-1}) f\|_2^2 < +\infty$. But $\Sigma_{n=1}^{\infty} n\|(E_n - E_{n-1}) f\|_2^2$ could easily diverge.

On the other hand, (*) holds for $p = 2$, as follows by the spectral theorem. In fact, if we write $Q^2 = \int_0^1 \lambda \, dE(\lambda)$, then $Q^{2n} = \int_0^1 \lambda^n dE(\lambda)$, and (*) becomes

$$\sum_{n=1}^{\infty} n \, \|(Q^{2n} - Q^{2n-2}) f\|_2^2 = \sum_{n=1}^{\infty} n \int_0^1 (\lambda^n - \lambda^{n-1})^2 (dE(\lambda) f, f)$$

$$\leq A_2 \|f\|_2^2 \ ,$$

which holds because

$$\sum_{n=1}^{\infty} n \int_0^1 (\lambda^n - \lambda^{n-1})^2 (dE(\lambda) f, f) = \int_0^1 \left[\sum_{n=1}^{\infty} n (\lambda^n - {}^{n-1})^2 \right] (dE(\lambda) f, f)$$

$$\leq A \|f\|_2^2$$

since $|\Sigma_{n=1}^{\infty} n (\lambda^n - \lambda^{n-1})^2| \leq A$ for $\lambda \in [0,1]$.

Thus, we see that the analogy between semigroups and martingales is by no means perfect. Let us proceed with the proof of (**). We have the inequality $\|(\Sigma_k |\Delta_k(f)|^2)^{1/2}\|_p \leq A_p \|f\|_p$ $(1 < p < +\infty)$, where

$$\Delta_k(f) = E_{2^k} f - E_{2^{k-1}} f \ ;$$

for this follows from Theorem 7 with $\{E_n\}$ replaced by $\{E_{2^n}\}$. (Note that it is *not* true that $(\Sigma_k |\Delta_k f|^2)^{1/2} \leq K(\Sigma_n |E_n f - E_{n-1} f|^2)^{1/2}$). Therefore, to prove (**), we need only show that

$$(***) \qquad \left\| \left(\sum_{n=1}^{\infty} n |\sigma_n - \sigma_{n-1}|^2 \right)^{1/2} \right\|_p \leq A_p \left\| \left(\sum_{k=1}^{\infty} |\Delta_k(f)|^2 \right)^{1/2} \right\|_p .$$

Recall that $S_n \equiv E_n f$, $a_n \equiv S_n - S_{n-1}$, and

$$\sigma_n \equiv \frac{S_0 + S_1 + \cdots + S_n}{n+1} = \sum_{k=0}^{n} \left(1 - \frac{k}{n+1}\right) a_k .$$

To prove (***) we study $\sigma_n - \sigma_{n-1}$. Write

$$(A) \quad \sigma_n - \sigma_{n-1} = \frac{-1}{n(n+1)} \sum_{k=0}^{n} k a_k = \frac{-1}{n(n+1)} \left[\sum_{j=0}^{[\log_2 n]} \sum_{k=2^j+1}^{2^{j+1}} k a_k + \sum_{k=<n>+1}^{n} k a_k \right],$$

where $<n>$ denotes the largest power of 2 which is smaller than n. Consider the inner sum, $\Sigma_{k=2^j+1}^{2^{j+1}} k a_k$. If the "k" in the product $k a_k$ were constant for each j, we could pull it out of the summation sign, and conclude that $\sigma_n - \sigma_{n-1}$ is essentially a sum of multiples of terms $\Sigma_{k=2^j+1}^{2^{j+1}} a_k = \Delta_{j+1}(f)$. This would reduce (***) to a very easy computation. Since k is not constant, we resort to summation by parts:

$$\sum_{2^j < k \leq 2^{j+1}} k a_k = 2^{j+1} \sum_{2^j < k \leq 2^{j+1}} a_k - \left[\sum_{2^j < k \leq 2^{j+1}-1} a_k + \left(\sum_{2^j < k \leq 2^{j+1}-2} a_k \right) + \cdots + (a_{2^j+1}) \right]$$

(B)

$$\equiv 2^{j+1} \Delta_{j+1}(f) - [E_{2^j-1}(\Delta_{j+1}(f)) + E_{2^{j+1}-2}(\Delta_{j+1}(f)) + \cdots + E_{2^j+1}(\Delta_{j+1}(f))]$$

by definition of E_ℓ, $\Delta_{j+1}(f)$, and a_k. This last equation is very encouraging—the first term on the right-hand side is precisely what $\Sigma_{2^j<k\leq 2^{j+1}} k a_k$ would be if k were constant ($k = 2^{j+1}$); and since there are about 2^{j+1} terms in the brackets, the sum in brackets should not be much larger than $2^{j+1}\Delta_{j+1}(f)$. We shall see in a moment that all these informal ideas can be made precise.

But first we need an analogue of equation (B), to handle the final sum in equation (A). The reader may easily check that

$$\text{(C)}\quad \sum_{k=\langle n\rangle+1}^{n} k a_k = n\cdot E_n(\Delta_{1+[\log_2 n]}(f)) - [E_{n-1}(\Delta_{1+[\log_2 n]}(f)) + E_{n-2}(\Delta_{1+[\log_2 n]}(f)) + \cdots + E_{\langle n\rangle+1}(\Delta_{1+[\log_2 n]}(f))] .$$

Putting (B) and (C) into (A), we obtain

$$\text{(D)}\quad \sigma_n - \sigma_{n-1} = \frac{1}{n(n+1)}\Bigg[-\sum_{j=0}^{[\log_2 n]} \sum_{2^j<k<2^{j+1}} [E_k(\Delta_{j+1}(f))] + \sum_{j=0}^{[\log_2 n]} 2^{j+1}\Delta_{j+1}(f) - \sum_{k=\langle n\rangle+1}^{n-1} E_k(\Delta_{1+[\log_2 n]}(f)) + n E_n \Delta_{1+[\log_2 n]}(f)\Bigg]$$

$$= \frac{-1}{n(n+1)}\Bigg[\sum_{k=0}^{n} \pm E_k(\Delta_{j(k)}(f)) + \sum_{j=0}^{[\log_2 n]} 2^{j+1}\Delta_{j+1}(f) + n E_n(\Delta_{1+[\log_2 n]}(f))\Bigg]$$

where $j(k)$ is an appropriate integer-valued function of k.

It is then a straightforward matter to use (D) together with Theorem 8 of Section 3 to obtain the inequality (***). The easy details are left to the reader.

The upshot of the last few paragraphs of estimates, is that (***) is actually an elementary consequence of some juggling with summation signs, and involves no deep ideas, except possibly Theorem 8.

Where do we stand? We have proved that

$$(**) \qquad \Big\| \Big(\sum_{n=1}^{\infty} n |\sigma_n - \sigma_{n-1}|^2 \Big)^{1/2} \Big\|_p \le A_p \|f\|_p \qquad (1 < p < +\infty)$$

where

$$\sigma_n = \left(\frac{E_0 + E_1 + \cdots + E_n}{n+1} \right) f$$

and the $\{E_k\}$ form a martingale or reverse martingale. Before we proceed to the proof of the Littlewood-Paley inequalities, let us prove the semi-group version of (**), which is:

$$(1) \qquad \Big\| \Big(\int_0^{\infty} t \Big| \frac{\partial}{\partial t} \sigma_t(f) \Big|^2 dt \Big)^{1/2} \Big\|_p \le A_p \|f\|_p \quad (1 < p < +\infty),$$

where $\{T^t\}$ is our semigroup, and $\sigma_t(f) \equiv 1/t \int_0^t T^s f\, ds$.

It is enough to prove (1) with $\int_0^\infty$ replaced by $\int_\epsilon^M$ $(0 < \epsilon < M < +\infty)$ and A_p independent of ϵ and M. For obviously we could then let $\epsilon \to 0+$, $M \to +\infty$. In the interval $[\epsilon, M]$, everything is real-analytic, so that we can replace all integrals in the right-hand side of (1) by Riemann sums, and all derivatives by finite difference quotients. These changes produce errors which tend to zero as the Riemann sums approach integrals, etc. So when all is said and done, (1) reduces the discrete analogue

$$(1') \qquad \Big\| \Big(\sum_{n=1}^{\infty} n |(\tilde{\sigma}_n - \tilde{\sigma}_{n-1})(f)|^2 \Big)^{1/2} \Big\|_p \le A_p \|f\|_p \qquad (1 < p < +\infty),$$

where $\tilde{\sigma}_n$ denotes $(T^0 + T^\epsilon + T^{2\epsilon} + \cdots + T^{n\epsilon})/n + 1$.

But (1′) is essentially trivial from (**) and Theorem 9. Let's plod through the proof. Theorem 9 immediately reduces (1′) to the inequality

$$(1'')\qquad \left\|\left(\sum_{n=1}^{\infty} n|\hat{E}(\sigma_n - \sigma_{n-1})|^2\right)^{1/2}\right\|_p \le A_p\|f\|_p ,$$

where the σ_n are as in (**), and $\hat{E}$ is some other conditional expectation operator. On the other hand,

$$\left\|\left(\sum_{n=1}^{\infty} n|\hat{E}(\sigma_n - \sigma_{n-1})|^2\right)^{1/2}\right\|_p \le \left\|\left(\sum_{n=1}^{\infty} n|\sigma_n - \sigma_{n-1}|^2\right)^{1/2}\right\|_p ,$$

simply because $\hat{E}$ has norm 1 on $L_p(\ell_2)$. (The proof that $\|\hat{E}\| = 1$ on $L_p(\ell_2)$ is the same as the proof that $\|\hat{E}\| = 1$ on L_p, and is left as an exercise for the reader.) Thus (1″) follows from (**). (1′) is proved.

This completes the proof of inequality (1). As we have seen, (1) is the strongest relevant conclusion we can possibly hope to squeeze out of martingale theory. At the same time, it is not strong enough to prove the Littlewood-Paley inequalities. So we have to resort to our familiar method of proving a stronger theorem for L_2, and then interpolating with (1).

To rephrase (1) in a form amenable to interpolation, we make use of the fractional integrals and fractional averages which served us so well in the first proof of the semigroup maximal theorem. In that proof, as the reader no doubt recalls, we set

$$M_a(f)(t) = \frac{t^{-a}}{\Gamma(a)} \int_0^t (t-s)^{a-1}\, T^s f\, ds$$

for $a \in \mathbf{C}$, $\operatorname{Re} a > 0$; and we saw that $M_a(f)$ could be continued analytically into the entire complex line $a \in \mathbf{C}$.

We can rewrite (1) in the form

$$\left\|\left(\int_0^{\infty} t\left|\frac{\partial}{\partial t} M_1(f)\right|^2 dt\right)^{1/2}\right\|_p \le A_p\|f\|_p .$$

By an elementary argument we shall deduce from this its analogue where M_1 is replaced by M_a, with $\operatorname{Re} a > 1$. In fact since $I_a f = I_{a-1} \cdot I_1 f$, then

$M_\alpha = t^{-\alpha} I_{\alpha-1}(tM_1)$.

Therefore

$$M_\alpha(f)(t) = \frac{t^{-\alpha}}{\Gamma(\alpha-1)} \int_0^t (t-s)^{\alpha-2} s M_1(f)(s)\,ds = \frac{1}{\Gamma(\alpha-1)} \int_0^1 (1-s)^{\alpha-2} s M_1(f)(st)\,ds .$$

So,

$$\frac{\partial M_\alpha}{\partial t}(f)(t) = \frac{1}{\Gamma(\alpha-1)} \int_0^1 (1-s)^{\alpha-2} s^2 M_1'(f)(st)\,ds .$$

Write now

$$\Phi_\beta(t) = \frac{1}{\Gamma(\beta)} \int_0^t (1-s)^{\beta-1} s^2 \phi(st)\,ds , \quad \mathrm{Re}\beta > 0 .$$

Then

$$\int_0^\infty |\Phi_\beta(t)|^2 t\,dt \le A_\beta^2 \int_0^\infty |\phi(t)|^2 t\,dt$$

with

$$A_\beta = \frac{1}{|\Gamma(\beta)|} \int_0^1 |(1-s)^{\beta-1}|\, s\,ds ,$$

which is at most $0(e^{\pi\|m\beta|})$ in any fixed strip lying strictly in the half-plane $\mathrm{Re}\beta > 0$.

Combining this with the above and (1) gives,

$$(2) \quad \left\|\left(\int_0^\infty t\left|\frac{\partial}{\partial t} M_\alpha(f)\right|^2 dt\right)^{1/2}\right\|_p \le A_{p,\alpha}\|f\|_p \quad (1 < p < +\infty) \quad (\alpha \in \mathbf{C},\ \mathrm{Re}\alpha > 1)$$

Inequality (2) would be the Littlewood-Paley inequality if we could set $\alpha = 0$, for in that case $M_\alpha(f) = T^t f$.

Next, we claim that if $p = 2$, then inequality (2) is valid for $\alpha = -1, -2, -3, \cdots, -k, \cdots$. We shall give the proof for $\alpha = -1$; the general case is no harder. $M_{-1}(f) = t(\partial/\partial t)T^t f$, so that

$$\frac{\partial}{\partial t} M_{-1}(f) = \frac{\partial}{\partial t}\left(t \frac{\partial}{\partial t} T^t f\right) = \frac{\partial}{\partial t} T^t f + t \frac{\partial^2}{\partial t^2} T^t .$$

Hence,

$$\left(\int_0^\infty t\left|\frac{\partial}{\partial t}M_{-1}(f)\right|^2 dt\right)^{1/2} \leq \left(\int_0^\infty t\left|\frac{\partial}{\partial t}T^t f\right|^2 dt\right)^{1/2} + \left(\int_0^\infty t^3\left|\frac{\partial^2}{\partial t^2}T^t f\right|^2 dt\right)^{1/2}$$

$$= g_1(f) + g_2(f),$$

where

$$g_k(f)(x) \equiv \left(\int_0^\infty t^{2k-1}\left|\frac{\partial^k}{\partial t^k}T^t f(x)\right|^2 dt\right)^{1/2} .$$

This shows that (2) holds for $p = 2$, $\alpha = -1$, since we already know from the proof of the semigroup maximal theorem that $f \to g_k(f)$ is a bounded operator on L_2.

Having proved an inequality for $\alpha = -1, -2, \cdots, -k, \cdots$ we can use the same argument as before, to deduce

$$\text{(3)} \qquad \left\|\left(\int_0^\infty t\left|\frac{\partial}{\partial t}M_\alpha(f)\right|^2 dt\right)^{1/2}\right\|_2 \leq A_\alpha \|f\|_2 \quad (\text{all } \alpha \in \mathbf{C}).$$

where A_α does not grow faster than $O(e^{\pi|\mathrm{Im}(\alpha)|})$, whenever α is restricted to a fixed strip of finite width, and $|\mathrm{Im}(\alpha)| \to \infty$.

Formally, it follows from (2) and (3) that

$$\left\|\left(\int_0^\infty t\left|\frac{\partial}{\partial t}M_\alpha(f)\right|^2 dt\right)^{1/2}\right\|_p \leq A'_{p\alpha}\|f\|_p \quad (1 < p < +\infty)$$

for all $\alpha \in \mathbf{C}$. In particular, as we have noted, the case $\alpha = 0$ is precisely the desired Littlewood-Paley inequality.

The trouble is of course, that the operators

$$f \to \left(\int_0^\infty t\left|\frac{\partial}{\partial t}M_\alpha(f)\right|^2 dt\right)^{1/2}$$

are non-linear, which means that we cannot apply the convexity theorem of Section 2, Chapter III in a totally simpleminded way. As in Chapter III, Section 3, the non-linearity is easy to overcome. Let $\Phi(t,x)$ be any measurable function on $(0,\infty)\times(M,\mathfrak{M},dx)$ such that $(\int_0^\infty t|\Phi(t,x)|^2 dt)^{1/2} \leq 1$ for every point x. The operators $T_\Phi^\alpha\colon L_p \to L_p$ defined by

$$T_\Phi^\alpha(f)(x) = \int_0^\infty t[\Phi(t,x)\cdot \frac{\partial}{\partial t} M_\alpha(f)(x,t)]dt$$

depend analytically on $\alpha \in C$. Inequality (2) implies that

$$\| T_\Phi^\alpha(f)\|_p \leq A_{p\alpha}\|f\|_p \qquad (1 < p < +\infty)$$

for $\operatorname{Re}\alpha > 1$ and inequality (1) implies that $\|T_\Phi^\alpha(f)\|_2 \leq A_\alpha\|f\|_2$ for all $\alpha \in C$. Since the T_Φ^α are linear operators, we can apply the convexity theorem to conclude that $\|T_\Phi^\alpha(f)\|_p \leq A'_{p\alpha}\|f\|_p$ $(1 < p < +\infty)$ for all $\alpha \in C$. Taking the sup over all Φ satisfying the defining condition, we obtain at last

$$\left\|\left(\int_0^\infty t\cdot\left|\frac{\partial}{\partial t}M_\alpha(f)\right|^2 dt\right)^{1/2}\right\|_p \leq A'_{p\alpha}\|f\|_p \qquad (1 < p < +\infty,\ \alpha \in C)\,.$$

Q.E.D.

Section 6. Dénouement

The following corollaries elaborate Theorem 10 and give an inkling of the applications of the Littlewood-Paley inequality.

COROLLARY 1. *For each* $k \geq 1$, $\|g_k(f)\|_p \leq A_p \|f\|_p$ $(1 < p < +\infty)$.

Proof: Proceed by induction on k. The case $k = 1$ is what we have just proved. We shall illustrate the induction step by considering $k = 2$; the general case is left to the reader.

As we saw above,

$$\frac{\partial}{\partial t} M_{-1}(f) = \frac{\partial}{\partial t} T^t f + t \frac{\partial^2}{\partial t^2} T^t f, \quad \text{i.e.,} \quad t^2 \frac{\partial^2}{\partial t^2} T^t f = \frac{\partial}{\partial t} M_{-1}(f) - \frac{\partial}{\partial t} T^t f .$$

Therefore

$$g_2(f) = \left(\int_0^\infty t^3 \left| \frac{\partial^2}{\partial t^2} T^t f \right|^2 dt\right)^{\frac{1}{2}} \leq \left(\int_0^\infty t \left| \frac{\partial}{\partial t} M_{-1}(f) \right|^2 dt\right)^{\frac{1}{2}} + \left(\int_0^\infty t \left| \frac{\partial}{\partial t} T^t f \right|^2 dt\right)^{\frac{1}{2}},$$

so that

$$\|g_2(f)\|_p \leq \left\| \left(\int_0^\infty t \left| \frac{\partial}{\partial t} M_{-1}(f) \right|^2 dt\right)^{\frac{1}{2}} \right\|_p + \|g_1(f)\|_p \leq A_p \|f\|_p$$

by inductive hypothesis and the last line of the proof of Theorem 10. Q.E.D.

COROLLARY 2 (Converse inequality). $\|f\|_p \leq A_p \|g_1(f)\|_p + \|E_0(f)\|_p$ $(1 < p < +\infty)$, *where* E_0 *is defined on* L_2 *as the orthogonal projection onto the space of all functions* h *such that* $T^t h = h$ *for* $t > 0$.

Proof: Here, E_0 plays the role of the constant term of a Fourier series. Note firstly that

$$(*) \qquad \|E_0(f)\|_p \leq A_p \|f\|_p \quad (1 < p < +\infty).$$

In fact, $T^t f \to E_0(f)$ in L_2 as $t \to +\infty$, for each $f \in L_2$. By our almost everywhere convergence theorems, $E_0(f) = \lim_{t\to\infty} T^t f \leq \sup_{t>0} |T^t f|$ pointwise. (*) now follows from the maximal theorem.

The corollary is now a consequence of Theorem 10 combined with the standard "isometry" arguments, based on the familiar identity

$$\|f\|_2^2 = 4\|g_1(f)\|_2^2 + \|E_0(f)\|_2^2$$

(see Section 3 in Chapter II). Q.E.D.

We can now formulate and prove a "Marcinkiewicz multiplier theorem" generalizing the Lie group version which appeared at the end of Chapter II. First of all, write $T^t = \int_0^\infty e^{-\lambda t} dE(\lambda)$, as is possible by the spectral theorem. For each bounded function $m(\lambda)$ on $(0, +\infty)$, set $T_m f \equiv \int_{0+}^\infty m(\lambda) dE(\lambda) f$. Thus T_m is a bounded operator on L_2.

COROLLARY 3. *Suppose that* m *is of Laplace transform type, i.e.,* $m(\lambda) \equiv \lambda \int_0^\infty e^{-\lambda t} M(t) dt$ $(\lambda > 0)$, *where* $M(t)$ *is a bounded function on* $(0, +\infty)$. *Then* T_m *is a bounded operator on all the spaces* L_p $(1 < p < +\infty)$.

Proof: The arguments we used to prove the Lie group version of this theorem in Chapter II apply to the present case, to prove that for $f \in L_p$, $g_1(T_m f) \leq K g_2(f)$ with K independent of f. Corollaries 1 and 2 now show that $\|T_m f\|_p \leq A_p \|f\|_p$. Q.E.D.

COROLLARY 4. *Let* A *be the infinitessimal generator of* $\{T^t\}$. *(Thus* $T^t = e^{tA}$.*) Then* $(-A)^{it}$ *is a bounded operator on* L_p $(1 < p < +\infty)$ *for each real* t.

Proof: λ^{it} is of Laplace-transform type. Q.E.D.

BIBLIOGRAPHICAL COMMENTS FOR CHAPTER IV

Section 1. For the theory of martingales see Doob [26], Chapter 7. The Marcinkiewicz interpolation can be found in Zygmund [20], Chapter XII.

Section 2. See the remarks after Theorem 7.

Section 3. Theorem 8 is new, as are Theorem 10 and its corollaries.

Section 4. For Theorem 9 see Rota [30] and Doob [27].

CHAPTER V

FURTHER ILLUSTRATIONS

In this final chapter we indicate some further illustrations of the theory, but our presentation is more in the spirit of Chapter II than Chapter III and Chapter IV.

Section 1. Lie groups

We assume that G is a non-compact, connected, Lie group. We let $X_1, X_2, \ldots, X_n$ be a basis for the (left-invariant) Lie algebra, considered as first-order differential operators on G. We set

$$\Delta^+ = \Sigma\ a_{ij} X_i X_j$$

where $\{a_{ij}\}$ is *any* real symmetric positive definite matrix. (More specific choices of the $\{a_{ij}\}$ will be made later.) Our first object is to consider the heat-diffusion semigroup $T_+^t = e^{t\Delta^+}$.

THEOREM. *There exists a semigroup* $\{T_+^t\}$, *which satisfies properties* (I), (II), (III) *and* (IV) *of Chapter III, and such that* $T_+^t = e^{t\Delta^+}$ *in the following two related senses:*

(a) $$\frac{(T_+^t - I)f}{t} \to \Delta^+ f, \quad \text{as} \quad t \to 0$$

for all sufficiently smooth f

(b) *If* $f \in L_p(G)$, $1 \leq p \leq \infty$, *then* $u(x,t) = (T_+^t f)(x) \in C^\infty(G \times R^+)$, *and*

$$\frac{\partial u(x,t)}{\partial t} = \Delta^+ u(x,t) .$$

It should be noted that the operators T_+^t are left-invariant, that is

$L_a T_+^t = T_+^t L_a$, where $(L_a f)(x) = f(a^{-1}x)$, $a \in G$.

According to Hunt's paper (see [12, p. 279]), we can construct a probability semigroup T_+^t which satisfies (a) and (III) ($T_+^t f \geq 0$, if $f \geq 0$) and (IV) ($T_+^t(1) = 1$). By that same construction, symmetry (our property (II)) is then also guaranteed by the symmetry of Δ^+ (see [12, Section 7.3]). (I) is a consequence of (II), (III), and (IV), and the Riesz convexity theorem. The fact that T_+ also satisfies conclusion (b) follows from the fact that

$$(T_+^t f)(x) = \int_G k_t^+(y) f(xy) dy$$

where k_t^+ is a fundamental solution, and $k_t^+ \in C^\infty(G \times R^+)$, also $k_t^+ \in L_1(G) \cap L_\infty(G)$. (For these facts see Nelson [14].)

As a simple corollary of the above we also obtain the existence of the "Poisson semigroup" corresponding to the equation

$$\frac{\partial^2 u}{\partial t^2}(x,t) + \Delta^+ u(x,t) = 0 .$$

In fact, define P_+^t by

$$P_+^t = \frac{1}{\sqrt{\pi}} \int_0^\infty \frac{e^{-u}}{\sqrt{u}} T_+^{t^2/4u} du$$

We then claim that $\{P_+^t\}$ is a semigroup which satisfies (I), (II), (III), (IV), and instead of (b) of the above theorem, we have

$$\frac{\partial^2 u}{\partial t^2} + \Delta^+ u = 0, \text{ where } u(x,t) = (P_+^t f)(x) .$$

The details of this passage from T_+^t to P_+^t can be carried out as in the analogous case of compact groups treated in Section 2 of Chapter II.

We now come to the g-function. For $f \in L_p(G)$ we consider the following two expressions

$$(1) \qquad \left(\int_0^\infty t \left| \frac{\partial (T_+^t f)}{\partial t}(x) \right|^2 dt \right)^{1/2}$$

(2) $$\left(\int_0^\infty t \left| \frac{\partial (P_+^t f)}{\partial t}(x) \right|^2 dt \right)^{1/2}$$

THEOREM 11. *Let* $f \in L_p(G)$, $1 < p < \infty$, *and let* $g_1(f)(x)$ *denote either of the two expressions above. Then*

$$B_p \|f\|_p \leq \|g_1(f)\|_p \leq A_p \|f\|_p \ .$$

This theorem is a direct consequence of Theorem 10 and its second corollary (Section 6 of the previous chapter), as soon as we verify that the projection E_0 is zero in this case. But suppose $f \in E_0(L_2(G))$. Then $T_+^t f = f$ for all $t \geq 0$, and so $f \in C^\infty(G)$, and $\Delta^+ f = 0$. Moreover since $k_t^+ \in L_2(G)$, for $t > 0$ and $f \in L_2(G)$, it follows that $f(x) = \int_G k_t^+(y) f(xy)\, dy$, vanishes at infinity. It may be assumed that f is real-valued. The above then shows that f attains its maximum and minimum values and thus must be constant and therefore zero in view of the local maximum principle of Hopf for the operator Δ^+. Notice also that $T_+^t f = f$ for all $t > 0$, if and only if $P_+^t f = f$ all t, and therefore the result applies also the the semi-group P_+^t.

The result for (2) can be extended when $1 < p \leq 2$ by taking into account the X_j derivatives. To do this define

$$(\mathbf{\Delta} u)(x,t) = \frac{\partial^2 u}{\partial t^2} + \Delta^+ u, \text{ and } |\nabla u|^2 = \left(\frac{\partial u}{\partial t}\right)^2 + \Sigma\, a_{ij}(X_i u)(X_j u) \ .$$

Let

$$g(f)(x) = \left(\int_0^\infty t\, |\nabla u|^2 dt \right)^{1/2}, \quad \text{if } u(x,t) = P_+^t(f)(x) \ .$$

Observe that $g_1(f)(x) \leq g(f)(x)$.

THEOREM 12.

$$\|g(f)\|_p \leq A_p \|f\|_p \ , \qquad 1 < p \leq 2 \ .$$

In proving this we follow closely the argument of Section 2 and 3 of

Chapter II. First, it suffices to consider the case where $f \geq 0$, and f is C^∞ and has compact support. Note that $k_t^+(x) > 0$, (see Nelson [14]). and thus $(T_+^t f)(x) > 0$, all $(x,t) \in G \times (0,\infty)$ and hence $(P_+^t f)(x) = u(x,t) > 0$, all $(x,t) \in G \times (0,\infty)$. Now it is immediate that

$$\text{(A)} \qquad \Delta u^p = p(p-1)u^{p-2}|\nabla u|^2 \ .$$

(See Lemma 2, Section 2 of Chapter II.)

Also

$$\text{(B)} \qquad \left(\int_G (\sup_{t>0} u(x,t))^p dx\right)^{1/p} \leq A_p\left(\int_G (f(x))^p dx\right)^{1/p}$$

by the general maximal theorem of Chapter III, or by the argument for the proof of Lemma 1 in Section 2, Chapter II. Finally,

$$\text{(C)} \qquad \int_0^\infty \int_G t(\Delta F)(x,t)\,dx\,dt = \int_G F(x,0)dx$$

for appropriate F defined in $G \times [0,\infty)$, and this class of F includes $F(x,t) = (u(x,t))^p$. The proof of (C) requires a little bit of care.

We have to observe first that if $f \in L_p(G)$, then

$$\int_G |u(x,t)|^p dx \to 0, \text{ as } t \to \infty, \text{ when } 1 < p < \infty ,$$

This assertion (not valid when $p = 1$) will be an immediate consequence of the Lebesgue dominated convergence theorem, the maximal theorem, and the fact that

$$\lim_{t \to \infty} u(x,t) = 0, \text{ for almost every } x ,$$

In proving the second assertion when $p = 2$, it suffices to consider a dense subset of f in $L_2(G)$. Now if $P_+^t = \int_0^\infty e^{-\lambda t} dE(\lambda)$, then we can write $f = \{f - E(\epsilon)f\} + E(\epsilon)f$ and $E(\epsilon)f \to 0$, as $\epsilon \to 0$, since we already saw that $E_0(f) = 0$. But when f_1 is of the form $f - E(\epsilon)f$, then $\|P_+^t f_1\| \leq e^{-\epsilon t}$. Moreover

$$\sup_{t>a} |P_+^t f_1| = \sup_{t>0} |P_+^t (P_+^a f_1)| \ .$$

Thus in view of the maximal theorem

$$\int_G \sup_{t>a} |P_+^t f_1|^2 dx \le A e^{-\epsilon 2a} \to 0, \text{ as } a \to \infty \ .$$

This shows that $\lim_{t\to\infty} u(x,t) = 0$ almost everywhere, for f in a dense subset of $L_2(G)$, and therefore all f in $L_2(G)$. Finally, $L_2(G) \cap L_p(G)$ is a dense subset of $L_p(G)$ and therefore $\lim_{t\to\infty} u(x,t) = 0$, almost everywhere for all $f \in L_p(G)$, and finally $\int_G |u(x,y)|^p dx \to 0$.

Let us return to the proof of (C) for all F of the form $(u(x,t))^p$. To establish this it will suffice, in view of what has just been done, to prove

$$\text{(C')} \qquad \int_0^N \int_G t(\Delta F)(x,t)dt = \int_G F(x,0)dx - \int_G F(x,N)dx \ .$$

Now if F, in addition to the smoothness it already has, also had as a function of x support in a fixed compact set of G, there would be no difficulty in verifying (C′) by the argument of integration by parts given in the proof of Lemma 3 (Section 2, Chapter II). To bring about this situation we construct a sequence $\{\phi_k(x)\}$ of C^∞ function G each with compact support, and the further properties that

(i) $\sup_k \sup_{x \in G} |P(X)\phi_k(x)| < \infty$, for any polynomial $P(X)$ of left-invariant differential operators;

(ii) $\phi_k(x) = 1$, for $x \in U_k$, where U_k are open sets with the property that $U_k \nearrow G$;

(iii) $\phi_{k+1}(x) \ge \phi_k(x) \ge 0$.

An example of such a sequence can be obtained as follows. Let $\eta(t)$, $0 \le t < \infty$ be a monotone C^∞ function in $(0,\infty)$, such that $\eta(t) = 1$ for t near zero and η vanishes outside a compact subset of t. Let $d(x)$ denote the square of the distance from $x \in G$ to the group identity, measured by any fixed smooth left-invariant Riemannian metric. Set $\phi_k(x) = \eta((d^2(x))/k)$.

Now with $F(x,t) = \phi_k(x)(u(x,t))^p$ the identity (C′) holds, as we have just indicated. We let $k \to \infty$, then the right side of (C′) clearly converges to $\int_G (u(x,0))^p dx - \int_G (u(x,N))^p dx$. The left-hand side of (C′) can be written as the sum of two integrals, whose integrands are respectively $-t(\Delta\phi_k)(u(x,t))^p$, and $t\phi_k \Delta (u)^p$. The first integral converges to zero since the $\Delta\phi_k$ are zero inside U_k, $U_k \to G$; and $|\Delta\phi_k| < A$, everywhere; also $u^p(x,t)$ is integrable on $G \times [0,N]$. The second integral converges monotonically to $\int_0^N \int_G t\Delta u^p dx\,dt$, since $\Delta u^p \geq 0$, and the ϕ_k converges monotonically to 1. This proves (C′) and therefore (C).

Now that (A), (B), and (C) are established the rest of the proof of Theorem 12 is then the same as the corresponding argument given in the compact case (for $1 < p \leq 2$) in Chapter II, (see Section 3 of that chapter).

It is important to point out that the argument for $p \geq 2$ given in the compact case cannot be extended in the present situation. This is because at that stage we would need to use the assertion that the X_j commute with P_+^t, which is the same as requiring that the X_j commute with $\Delta^+ = \Sigma\, a_{ij} X_i X_j$. For similar reaons some of the further applications given in Chapter II for compact groups do not have evident analogies in the case of general non-compact G, but there seem to be interesting possibilities if we make the assumption that G is semi-simple as we shall now see.

Section 2. Semi-simple case

We now assume that G is a unimodular Lie group, K is a compact subgroup, and we consider the homogeneous space $S = G/K$.

As usual $L_p(S, ds)$, (where ds is G invariant measure on G/K) is identifiable with the class of functions $\{f \mid f \in L_p(G)$, and $f(gk) = f(g)$, $k \in K\}$. We also make a more specific choice of the left-invariant Laplacian $\Delta^+ = \Sigma\, a_{ij} X_i X_j$, by requiring that Δ^+ is also right-invariant under the action of K. More particularly if we write $\rho_k(f)(x) = f(xk)$, then we require that $\Delta^+ \rho_k f = \rho_k \Delta^+ f$, $k \in K$, for all sufficiently smooth functions f on G. We can obtain such a positive definite symmetric matrix $\{a_{ij}\}$, by starting with any positive definite symmetric matrix $\{a_{ij}^{(o)}\}$ and performing the appropriate integration with respect to the compact group K; (see the argument in Section 7 of Chapter I). When we have such a Δ^+ which is right-invariant under K, then we denote by Δ its induced action on functions on S.

Let us denote by 0 the origin in S, that is, the point corresponding to the coset K. Then our non-unique choice of $\{a_{ij}\}$ corresponds to a choice of a positive definite quadratic form in the tangent space at 0, invariant under the action of K. For every such quadratic form we get a Riemannian metric on S, invariant under the action of G, and Δ then is the Laplace-Beltrami operator for this metric on S (see the related problem at the end of Section 7, Chapter I).

By the construction given in Section 1 above, the operator Δ leads to semigroups which we now write as T^t and P^t (instead of T^t_+ and P^t_+); since the latter semigroups are right-invariant under K, the former semigroups act on $L_p(S, ds)$. It also follows that P^t and T^t satisfy properties (I), (II), (III), and (IV), our fundamental properties for symmetric diffusion semi-groups. We can write symbolically

$$T^t = e^{t\Delta}, \text{ and } P^t = e^{-t(-\Delta)^{1/2}} .$$

In addition if $U(x,t) = (T^t f)(x)$, and $u(x,t) = P^t(f)(x)$, then $\partial U/\partial t = \Delta U(x,t)$, and $(\partial^2 u/\partial t^2) + \Delta u(x,t) = 0$.

There are now certain theorems for $L_p(S, ds)$ which are immediate consequences of the corresponding results for G (Theorems 11 and 12) in the previous section. We need not reformulate these theorems separately.

We now make the assumption that $S = G/K$ is a symmetric space. If $\underline{g}$ is the Lie algebra of G, and $\underline{k}$ the sub-algebra corresponding to K, we have that $\underline{g} = \underline{k} + \underline{p}$ where $\underline{p}$ is the orthogonal complement of $\underline{k}$ in $\underline{g}$ with respect to the Killing form. Let now $\tilde{X}_1, \tilde{X}_2, \ldots, \tilde{X}_n$ be a basis of the *right*-invariant Lie algebra $\underline{g}$ so that $\tilde{X}_1, \ldots, \tilde{X}_r$ is a basis for $\underline{k}$ and $\tilde{X}_{r+1}, \tilde{X}_{r+2}, \ldots, \tilde{X}_r$ is a basis for $\underline{p}$. Then there exists two positive definite symmetric matrices $\{c_{ij}\}$, $1 \leq i, j \leq r$, and $\{b_{k\ell}\}$, $r+1 \leq k, \ell \leq n$, so that

$$\Delta^- = \sum_{r+1\leq k,\ell\leq n} b_{k\ell}\tilde{X}_k\tilde{X}_\ell - \sum_{1\leq i,j\leq r} c_{ij}\tilde{X}_i\tilde{X}_j$$

is not only right-invariant, but also left-invariant. (Δ^- in effect is the Casimir operator: see Helgason [2, p. 451]).

Let us say that a function f is *zonal* if $f(k_1 x k_2) = f(x)$, where k_1, $k_2 \in K$. These are exactly the functions on $S = G/K$, which are also invariant under the action of K on S. It is to be noted that if f is any smooth zonal function, then

$$\sum_{1\leq i,j\leq r} c_{ij}\tilde{X}_i\tilde{X}_j \quad f \equiv 0 \ ,$$

and therefore $\Delta^- f$ is a Laplace-Beltrami operator of f, which is identical with Δf for appropriate a_{ij}. We fix this choice of a_{ij} in what follows. Notice also that if

$$|\nabla f|^2 = \sum_{1\leq i,j\leq n} a_{ij}(X_i f)(X_j f),$$

then if f is zonal we have

$$|\nabla f|^2 = \sum b_{ij}(\tilde{X}_i f)(\tilde{X}_j f)$$

We can now state the version of Theorem 2 in Section 2, Chapter II, valid for all p, $1 < p < \infty$, but for zonal f.

THEOREM 13. *Let* $f \in L_p(S)$, *and assume that* f *is zonal. Let* $u(x,t) = P^t(f)(x)$. *Set*

$$g(f)(x) = \left(\int_0^\infty t\,|\boldsymbol{\nabla} u|^2 dt\right)^{1/2},$$

where

$$|\boldsymbol{\nabla} u|^2 = \left(\frac{\partial u}{\partial t}\right)^2 + |\nabla u|^2.$$

Then for $1 < p < \infty$

$$B_p\|f\|_p \leq \|g(f)\|_p \leq A_p\|f\|_p \ .$$

The theorem is proved as follows. Since $g_1 \leq g$, the inequality $B_p\|f\|_p \leq \|g(f)\|_p$ follows directly from Theorem 11 in the previous section. In addition the case $p \leq 2$ of the inequality $\|g(f)\|_p \leq A_p\|f\|_p$, is contained in Theorem 12. It remains to consider the case $p > 2$ of the direct inequality.

To do this we follow closely the argument given in Theorem 2 of Section 2, Chapter II, in particular "part II" of that proof. The basic step was to prove the inequality

$$\int_G (g(f)(x))^2\phi(x)dx = \int_0^\infty \int t|\nabla u(x,t)|^2\phi(x)dx\,dt$$

$$\leq A\int_0^\infty \int_G t|\nabla u(x,t)|^2\phi(x,t)dx\,dt \qquad (A = 4),$$

if $\phi \geq 0$, and $\phi(x,t) = P^t(\phi)(x)$. (Here ϕ is also zonal.)

In the compact case this can be proved, as is pointed out in Chapter II, because the X_i commute with P^t. In the present case the $\tilde{X}_i$ being right-invariant commute with the P^t, the latter being left-invariant, and of course as we have already remarked $|\nabla u(x,t)|^2$ can be expressed in terms of the $\tilde{X}_i u(x,t)$. The rest of the proof of the theorem is quite parallel with that of the compact case and may be left as an exercise to the interested reader.

We now come to the analogue of the Riesz transforms (see e.g., Section 4 of Chapter II), in the present case of zonal functions for symmetric spaces.

We prove first the inequality

$$(*) \qquad \sum_i \|\tilde{X}_i P^t(f)\|_p \leq A_p\left\|\frac{\partial P^t}{\partial t}(f)\right\|_p, \quad t > 0, \quad 1 < p < \infty,$$

with A_p independent of t. Fix a $t = t_0$, and assume that to begin with f is C^∞ and has compact support. Then $(\tilde{X}_i P^{t_0})(f) = P^{t_0}(\tilde{X}_i f)$ and the left side of (*) is well-defined. The right side of (*) is of course well-defined on general principles in view of the fact that $t \to P^t(f)$ is analytic in t (as a function with values in L_p) in a proper sector which contains the positive t semi-axis, when $1 < p < \infty$; see Theorem 2 of Chapter III. Now let $f_i = \tilde{X}_i P^{t_0}(f) = P^{t_0}(\tilde{X}_i(f))$, and $f_0 = (\partial P^t/\partial t)(f)|_{t=t_0}$.

Bt Theorem 11 in the previous section

$$\|f_i\|_p \leq B_p^{-1} \left\| \left(\int_0^\infty t \left| \frac{\partial P^t(f_i)}{\partial t} \right|^2 dt \right) \right\|_p$$

and because of Theorem 13 we have

$$\left\| \left(\int_0^\infty t |\tilde{X}_i (P^t f_0)|^2 dt \right)^{1/2} \right\|_p \leq A_p \|f_0\|_p \quad .$$

Together this gives

$$(**) \qquad \sum \|f_i\|_p \leq C_p \|f_0\|_p$$

which is (*) for smooth f with compact support. The passage of (*) and (**) to general f is then by a routine limiting argument.

The definition of the Riesz transforms can be given symbolically as

$$R_i(f) = \tilde{X}_i (-\Delta)^{-1/2} f \; .$$

In order to give a precise definition we proceed as follows. Consider the f_0 of the form

$$f_0 = \frac{\partial P^t}{\partial t}(f) \Big|_{t=t_0} \quad \text{where } t_0 > 0 \text{ and } f \in L_p \cap L_2 \quad .$$

For these f_0 we define the Riesz transforms by $R_i(f_0) = -f_i = -\tilde{X}_i P^{t_0}(f)$ in accordance with the inequality (*) or (**). Now purely formally since $P^t = e^{-t(-\Delta)^{1/2}}$ we have $f_0 = -(-\Delta)^{1/2} P^{t_0}(f)$, and therefore $-f_i = \tilde{X}_i P^{t_0}(f)$ $= \tilde{X}_i(-\Delta)^{-1/2} f_0$.

In order to show that these R_i are in fact well-defined or a dense subset of L_p, we need to observe the following two simple facts:

(i) The set of f_0 of the form

$$f_0 = \frac{\partial P^t}{\partial t}(f)\Big|_{t=t_0} \quad (\text{for some } t_0 > 0, \text{ with } f \in L_p \cap L_2),$$

is dense in L_p $1 < p < \infty$. To see this, recall that for any

$$h \in L_2 \cap L_p\,, \quad \lim_{\substack{t_0 \to 0 \\ t \to \infty}} P^{t_0}h - P^{t'}h = h$$

in both L_p and L_2 norm (when $1 < p < \infty$), as is shown in the proof of Theorem 12 of the previous section. Thus the set of f_0 of the form $f_0 = (P^{t_0} - P^{t'})h$, $h \in L_2 \cap L_p$ with $t' > t_0$, is dense in L_p. Each such f_0 can be represented in the form

$$f_0 = \frac{\partial P^t(f)}{\partial t}\Big|_{t = t_0} \quad \text{where} \quad f = -\int_0^{t'-t_0} P^t(h)dt\ ,$$

as an easy calculation verifies

(ii) To see that the resulting $R_i(f_0)$ is well-defined remark the following. Suppose f_1 and $f_2 \in L_p$, and

$$f_0 = \frac{\partial P^t}{\partial t}(f_1)\Big|_{t=t_1} = \frac{\partial P^t}{\partial t}(f_2)\Big|_{t=t_2}, \quad t_1, t_2 > 0.$$

Then $P^{t_1}(f_1) = P^{t_2}(f_2)$. This is because

$$P^{t_j}(f) = -\lim_{t' \to \infty} \int_{t_j}^{t'} \frac{\partial P^t}{\partial t}(f)dt$$

in the L_p norm, since $P^{t'}(f) \to 0$, as $t' \to \infty$. However by the semigroup property $P^t(P^{t_j}f) = P^{t+t_j}(f)$ and therefore

$$P^t\left(\left(\frac{\partial P^t}{\partial t}(f)\right)\Big|_{t=t_j}\right) = \frac{\partial P^{t+t_j}}{\partial t}(f)\,,$$

from which our desired conclusion follows.

The way the $R_i(f)$ have been defined shows that by (*) (or (**)) we have

$$\|R_i(f)\|_p \leq A_p\|f\|_p$$

for a dense linear subset in L_p, $1 < p < \infty$ and hence the R_i have a unique bounded extension to all of L_p. We summarize this result and elaborate it somewhat as follows.

THEOREM 14. *Suppose* $1 < p < \infty$, *and* f *is zonal* $f \in L_p$. *Then:*

(a) $B_p\|f\|_p \leq \Sigma_i \|R_i(f)\|_p \leq A_p\|f\|_p$.

(b) *When* $p = 2$,

$$\sum_{i,j} b_{ij} \int_S R_i(f)R_j(f)ds = \int_S (f)^2 ds \ .$$

The proof of the inequality $\Sigma\|R_i(f)\|_p \leq A_p\|f\|_p$ has been given above, and this shows in particular that it suffices to prove (b) for a class of f which are dense in L_2.

Start with f_0 which is C^∞ and has compact support and set $f_1 = P^{t_1}(f_0)$, $t_1 > 0$, $u(x,t) = P^{t+t_1}(f_0) = P^t(f_1)$. Apply the identity (C) (of the proof of Theorem 12), wtih $F = u^2$. Then since $\Delta(u^2) = 2|\nabla u|^2$, and if f_0 is zonal, so is f_1 and $u(x,t)$, we have

$$\Delta u^2 = 2[[\frac{\partial u}{\partial t}]^2 + \sum_{i,j} b_{ij}\tilde{X}_i(u)\tilde{X}_j(u)] \ .$$

$$\int_S |u(x,t_1)|^2 ds = 2\int_0^\infty \int_S t\Big[\Big|\frac{\partial P^{t+t_1}}{\partial t}(f_0)\Big|^2$$

$$+ \sum b_{ij}\tilde{X}_i(P^{t+t_1}f_0)\tilde{X}_j(P^{t+t_1}f_0)\Big] ds\,dt \ .$$

Now the identity

$$\|f_1\|_2^2 = 4\|g_1(f_1)\|_2^2 + \|E_0(f_1)\|_2^2$$

is valid for any one of our general semigroups, and follows easily by the spectral representation of P^+, which can be written as

$$P^t = E_0 + \int_{0+}^{\infty} e^{-\lambda t} dE(\lambda)$$

for appropriate $E(\lambda)$.

We have, in fact, already pointed out that in the present case $E_0 = 0$. We therefore have

$$\int_S |u(x,t_1)|^2 ds = 4\int_0^{\infty}\int_S t\left|\frac{\partial P^{t+t_1}}{\partial t}\right|^2 ds\,dt .$$

Hence

$$\int_0^{\infty}\int_S t\left|\frac{\partial P^{t+t_1}}{\partial t}(f_0)\right|^2 ds\,dt = \int_0^{\infty}\int_S t \sum_{i,j} b_{ij}\tilde{X}_i P^{t+t_1}(f_0)\tilde{X}_j P^{t+t_1}(f_0) ds\,dt.$$

Since this is true for every t_1 and $\tilde{X}_i P^{t+t_1} = P^{t+t_1}\tilde{X}_i$ we obtain

$$\int_S \left|\frac{\partial P^t}{\partial t}(f_0)\right|^2 ds = \int_S \sum b_{ij}\tilde{X}_i P^t(f_0)\tilde{X}_j P^t(f_0) ds$$

for all $t > 0$.

This identity is (b) for $f = (\partial P^t/\partial t)(f_0)$, $t > 0$ and since this class of f is easily seen to be dense in L_2 the identity (b) is then fully proved. By polarization this identity yields, for $f, g \in L_2$

$$\int_S fg\,ds = \int_S \sum b_{ij} R_i(f) R_j(g) ds$$

and hence by Hölder's inequality

$$\left|\int_S fg\,ds\right| \leq C\left(\sum_i \|R_i(f)\|_p\right)\left(\sum_i \|R_i(g)\|_q\right)$$

where $1/p + 1/q = 1$, wherever $f \in L_2 \cap L_p$, and $g \in L_2 \cap L_q$. However,

$$\|f\|_p = \sup_{\substack{g \in L_2 \cap L_q \\ \|g\|_q \leq 1}} \left|\int_S fg\,ds\right| .$$

Thus

$$\|f\|_p \leq B_p^{-1} \sum_i \|R_i(f)\|_p$$

by the inequality $\| \Sigma R_i(g)\|_q \leq A_q \|g\|_q$ which we already know. Hence we have shown $B_p \|f\|_p \leq \Sigma \|R_i(f)\|_p$, whenever $f \in L_2 \cap L_p$ and a final trivial limiting argument removes the restriction that $f \in L_2$.

*Section 3. Sturm-Liouville**

Let L be a Sturm-Liouville operator defined on an interval (a_1, a_2) of the line. Thus

$$L(f) = a(x)\frac{d^2f}{dx} + b(x)\frac{df}{dx} + c(x)f$$

where it is assumed that a, b and c are continuous and $a(x) > 0$, while $c(x) \leq 0$. Then we can find an appropriate measure $q(x)dx$ on (a_1, a_2) so that with respect to $L_2(q(x)dx)$, L is formally self-adjoint. With the appropriate boundary conditions the semigroup $T^t = e^{tL}$ can be constructed and satisfies the general conditions (I), (II), and (III) of Chapter III. Moreover if $c(x) \equiv 0$ then T^t also satisfies (IV) (see McKean, [13]).

These semigroups therefore give us interesting examples for which the results of Chapters III and IV apply.

It will be the purpose of the concluding remarks that follow to follow up the hint that part of the more refined analysis—carried out in Chapter II and in the present chapter in the context of groups—is most probably valid also in the setting of Sturm-Liouville expansions.

We shall be suggesting only the statements and proofs of certain facts that are indicated by our experience with the theory for compact and semisimple groups and the results in [37] for other classical expansions. The reader who is wary of this kind of speculation need not pursue this matter further.

---

*See Appendix (1985).

Section 4. Heuristics

It will be convenient to rewrite L in a more special form,

$$L(f) = \frac{d^2f}{dx^2} + a(x)\frac{df}{dx} \quad ,$$

(to which the general L may anyway be reduced by changes of variable).

This L is formally self-adjoint with respect to the measure $q(x)dx$, where $(q'(x))/(q(x)) = a(x)$, i.e., $q(x) = e^{\int a(x)dx}$.

In an important part of what follows it will be necessary to assume that $a'(x) \leq 0$. This is indeed the case for many interesting classical expansions. For example $a(x) = 2\lambda/x$, then $q(x)dx = x^{2\lambda}dx$, $0 < x < \infty$ which corresponds to the Fourier-Bessel and ultra-spherical expansions (see [37]). Also if $a(x) = -2x$, $q(x)dx = e^{-x^2}dx$, $-\infty < x < \infty$ which gives rise to the Hermite expansion.

We consider next the semigroup $T^t = e^{tL}$, once we have imposed appropriate boundary conditions, and as indicated before it satisfies all the conditions (I), (II), (III), and (IV) of a symmetric diffusion semigroup. We then consider the corresponding sub-ordinated semigroup $P^t = e^{-t(-L)^{1/2}}$ which can be expressed as

$$P^t = \frac{1}{\sqrt{\pi}} \int_0^\infty \frac{e^{-u}}{\sqrt{u}} T^{t^2/4u} du \; .$$

Then if $u(x, t) = (P^t f)(x)$, u satisfies the equation $\Delta u(x,t) = 0$, where

$$\Delta = \frac{\partial^2}{\partial x^2} + a(x)\frac{\partial}{\partial x} + \frac{\partial^2}{\partial t^2} \quad .$$

We write

$$g(f)(x) = \left(\int_0^\infty t\left(\left|\frac{\partial u}{\partial t}\right|^2 + \left|\frac{\partial u}{\partial x}\right|^2\right)dt\right)^{1/2} \quad ,$$

and $\|f\|_p = (\int |f(x)|^p q(x)dx)^{1/p}$.

The first conjectural theorem is the statement

H, 1: $$\|g(f)\|_p \leq A_p\|f\|_p \,, \qquad 1 < p < \infty$$

and conversely $\|f\|_p \leq A_p\|g(f)\|_p$, *whenever* f *is orthogonal to the eigenfunction of* L *of eigenvalue zero, (if such exists).*

The proof should follow closely the pattern of Theorem 2 of Chapter II and Theorem 13 of the present chapter.

To begin with, when $p \leq 2$, the assumption $a'(x) \leq 0$ seems unnecessary. We set down the following facts

(A′) $$\Delta u^p = p(p-1)u^{p-2}\left[\left(\frac{\partial u}{\partial x}\right)^2 + \left(\frac{\partial u}{\partial t}\right)^2\right]$$

$$\text{if } u > 0,\ u(x,t) = P^t(f)(x) \,.$$

(B′) $$\int \left|\sup_{t>0} u(x,t)\right|^p q(x)dx \leq A_p \int |f(x)|^p q(x)dx$$

(C′) $$\int_0^\infty \int t\Delta F(x,t)q(x)dx\,dt = \int F(x,0)\,q(x)dx - \int F(x,\infty)\,q(x)dx$$

and try to apply (C′) when $F = (u(x,t))^p$.

Only the argument involving (C′) would seem to require further analysis, since (B′) is a consequence of the general maximal theorem of Chapter III and (A′) is merely straightforward differentiation.

To pass to the case $p \geq 2$ (here we would require the condition $a'(x) \leq 0$) we would need the assertion that

$$\int_0^\infty \int t\left(\left(\frac{\partial u}{\partial t}\right)^2 + \left(\frac{\partial u}{\partial x}\right)^2\right)\phi(x)q(x)dx\,dt$$

$$\leq A \int_0^\infty \int t\left(\left(\frac{\partial u}{\partial t}\right)^2 + \left(\frac{\partial u}{\partial t}\right)^2\right)\phi(x,t)q(x)dx\,dt$$

for $\phi(x) \geq 0$, with $\phi(x,t) = P^t(\phi)(x)$. ($A = 4$.)

This can be obtained (at least formally) on the basis of a "sub-harmonic" property of $(\partial u/\partial t)^2 + (\partial u/\partial x)^2$. In fact we claim that we always have

$$\Delta\left[\left(\frac{\partial u}{\partial t}\right)^2 + \left(\frac{\partial u}{\partial x}\right)^2\right] \geq 0, \text{ if } u \text{ satisfies } \Delta u = 0.$$

Write $U = \partial u/\partial t$, and $V = \partial u/\partial x$. Then clearly $\Delta U = 0$, and also $\dot{\Delta}(V) = \Delta(V) + a'(x)V = 0$. Next $\Delta(U^2) = 2(U_x^2 + U_t^2) + 2U\Delta(U)$ and also $\Delta(V^2) = 2(V_x^2 + V_t^2) + 2V\dot{\Delta}(V)$. Here $U_x = \partial U/\partial x$, $U_t = \partial U/\partial t$, etc.

Together then $\Delta(U^2 + V^2) = \text{positive} + 2V\Delta V = \text{positive} - 2a'(x)V^2$, which is ≥ 0, if $a'(x) \leq 0$.

The converse inequality $A_p^{-1}\|f\|_p \leq \|g(f)\|_p$ for all f orthogonal to the eigenfunction of eigenvalue zero is of course a consequence of the general result in Section 5 of Chapter IV, and the fact that $g_1(f) \leq g(f)$.

Another converse inequality for f satisfying the same conditions is concontained in the following:

H, 2: $$\|f\|_p \leq A_p\|g_x(f)\|_p \,, \qquad 1 < p < \infty$$

where $g_x(f) = \left(\int_0^\infty t(\partial u/\partial x)^2 dt\right)^{1/2}$, $u(x,t) = P^t(f)(x)$.

In fact, it can be shown without difficulty, using (A′) and the L_2 theorem for g_1, that $4\|g_x(f)\|_2^2 = \|f\|_2^2 - \|E_0(f)\|_2^2$ where E_0 is the projection corresponding to the eigenfunction of eigenvalue 0. Thus, if $E_0(f) = 0$, we get by polarization that

$$\int ff_1\, q(x)dx = \int\int \left(\frac{\partial P^t}{\partial x}(f)(x)\right)\left(\frac{\partial P^t}{\partial x}(f_1)(x)\, t\, dt\, q(x)dx\right)$$

and thus by Hölder's inequality and the direct part of H, 1 we get H, 2.

The next idea which merits some further examination is the possibility of studying the notion of the Hilbert transform (i.e., Riesz transform) in this context.

To illustrate this idea, assume for the moment that L has a discrete spectrum: that is, suppose that $\phi_1, \phi_2, \ldots, \phi_k, \ldots$ form a complete set of eigenfunctions with eigenvalues $-\lambda_1^2, \ldots, -\lambda_k^2, \ldots$. Assume therefore that $L\phi_k + \lambda_k^2\phi_k = 0$, and $\int \phi_k(x)\phi_\ell(x)q(x)dx = \delta_{k,\ell}$.

Then a formal integration by parts shows that the functions

$$\left\{\frac{1}{\lambda_k}\frac{d\phi_k}{dx}\right\}$$

form an orthonormal set again, that is

$$\int \frac{d\phi_k}{dx} \frac{d\phi_\ell}{dx} q(x)dx = \lambda_k^2 \delta_{k,\ell} .$$

The proposed Hilbert transform would provide the mapping

$$\phi_k(x) \to \frac{1}{\lambda_k} \frac{d\phi_k}{dx} .$$

These observations can be put in a general setting by observing that if the ϕ_k are eigenfunctions of $L \equiv (d^2/dx^2) + a(x)(d/dx)$ then the $d\phi_k/dx$ are eigenfunctions of $\tilde{L} \equiv (d/dx^2) + a(x)(d/dx) + a'(x)$.

We recall the general assumption $a'(x) \leq 0$ that we made earlier. It ensures the existence of a semigroup $\tilde{T}^t = e^{t\tilde{L}}$, which satisfies our general conditions (I), (II), (III), (after appropriate boundary conditions have been supplied).

Let now $\tilde{P}^t$ be the corresponding sub-ordinated semigroup corresponding to the differential equation $\tilde{\Delta} u(x,t) \equiv$

$$\tilde{\Delta} u(x,t) \equiv \frac{\partial^2 u}{\partial x^2} + a(x)\frac{\partial u}{\partial x} + a'(x)u + \frac{\partial^2 u}{\partial t^2} = 0 .$$

We define

$$\tilde{g}(f)(x) = \left(\int_0^\infty t \left[(\frac{\partial u}{\partial t})^2 + (\frac{\partial u}{\partial x})^2 - a'(x)u^2 \right] dt \right)^{1/2}$$

and we can expect that:

H,3: $$\|\tilde{g}(f)\|_p \leq A_p \|f\|_p , \qquad 1 < p \leq 2 .$$

The indicated proof of this would follow the same pattern (A′), (B′), (C′) as outlined for the operator L, except that the identity for (A′) would now read

$$\Delta u^p = pu^{p-2}[(p-1)((\frac{\partial u}{\partial t})^2 + (\frac{\partial u}{\partial x})^2 - a'(x)u^2]$$

if $\tilde{\Delta} u = 0$, and $u > 0$.

Observe that if we set

$$u(x,t) = \frac{\phi_k(x)e^{-\lambda_k t}}{\lambda_k}$$

then $\Delta u = 0$, and $-(\partial u/\partial t)(x,0)$ and $(\partial u/\partial x)(x,0)$ are Hilbert transforms in the indicated sense.

The theorem that would seem to emerge from these considerations can be stated as follows:

H, 4: *Let* $f \in L_p(q(x)dx)$ *and* $u(x,t) = P^t(f)(x)$. *Then*

$$\|\frac{\partial u}{\partial t}(x,t)\|_p \leq A_p \|\frac{\partial u}{\partial x}(x,t)\|_p, \quad t > 0, \quad 1 < p \leq 2$$

with A_p *independent of* f *or* t.

Since

$$\frac{\partial u}{\partial t}(x,t+t_1) = P^t\left(\frac{\partial u}{\partial t}(x,t_1)\right)$$

on application of H, 2 shows that

$$\|\frac{\partial u}{\partial t}(x,t_1)\|_p \leq A_p \left\|\left(\int t \left|\frac{\partial^2 u}{\partial x \partial t}(x,t+t_1)\right|^2 dt\right)^{1/2}\right\|_p .$$

Next

$$\frac{\partial u}{\partial x}(x,t+t_1) = \tilde{P}^t\left(\frac{\partial u}{\partial x}(x,t_1)\right)$$

and therefore by H, 3,

$$\left\|\left(\int_0^\infty t \left|\frac{\partial^2 u}{\partial x \partial t}(x,t+t_1)\right|^2 dt\right)^{1/2}\right\|_p \leq A_p \|\frac{\partial u}{\partial x}(x,t_1)\|_p .$$

Combining these two gives H, 4.

There are also inequalities of this type for $p \geq 2$, but we shall not here pursue this matter further.

BIBLIOGRAPHICAL COMMENTS FOR CHAPTER III

Section 5. The cases corresponding to Hermite and Laguerre polynomials have recently been treated by Muckenhoupt [38].

REFERENCES

(LIE) GROUPS

[1] C. Chevalley, "Theory of Lie Groups" Vol. 1, Princeton, N.J., 1946.

[2] S. Helgason, "Differential Geometry and Symmetric Spaces," New York, 1962.

[3] L. H. Loomis, "An introduction to Abstract Harmonic Analysis," New York, 1953.

[4] K. Nomizu, "Lie Groups and Differential Geometry," (Math. Soc, Japan Publ., Vol. 2) 1956.

[5] R. Palais et al, "Seminar on the Atiyah-Singer Index Theorem," Princeton, 1965.

[6] L. S. Pontrjagin, "Topological Groups," Princeton, 1939.

[7] ______, 2nd edition, Moscow, 1954, English transl., New York, 1966.

[8] A. Weil, "L'intégration dans les groups topologiques et les applications," Hermann, Paris, 1940.

SEMIGROUPS OF OPERATORS

[9] N. Dunford and J. T. Schwartz, "Linear Operators," Part I, New York, 1958.

[10] W. Feller, "An Introduction to Probability Theory and its Applications," New York, 1966.

[11] E. Hille and R. S. Phillips, "Functional Analysis and Semi-groups," Amer. Math. So., 1957.

[12] G. Hunt, Semi-groups of Measures on Lie Groups, Trans. Amer. Math. Soc., 81 (1956), 264-293.

[13] H. P. McKean, Jr., Elementary Solutions for Certain Parabolic Partial Differential Equations, Trans. Amer. Math. Soc., 82 (1956), 519-548.

[14] E. Nelson, Analytic Vectors, Ann. of Math., 70 (1959), 572-615.

[15] R. S. Phillips, On the Integration of the Diffusion Equation with Boundary Conditions, Trans Amer. Math. Soc., 98 (1961), 62-84.

LITTLEWOOD-PALEY THEORY

[16] J. E. Littlewood and R. E. Paley, Theorems on Fourier Series and Power Series, (I), Jour. London Math. Soc., 6 (1931), 230-3; (II) Proc. London Math. Soc., 42 (1936), 52-89; (III) *ibid.*, 43 (1937), 105-26.

[17] R. E. Paley, ' Remarkable Series of Orthogonal Functions, I, Proc. London Math. Soc., 34 (1932), 241-79.

[18] E. M. Stein, On the Functions of Littlewood-Paley, Lusin, and Marcinkiewicz, Trans. Amer. Math. Soc., 88 (1958), 430-466.

[19] _____, "Intégrals Singulieres et fonctions differentiables de plusieurs variables," Lecture notes at Faculté des Sciences, Orsay, France; academic year 1966-67.

[20] A. Zygmund, "Trigonometric Series," Second edition, Cambridge Univ. Press, 1959.

INTERPOLATION OF OPERATORS

[21] A. Benedeck and R. Panzone, The Spaces L^P With Mixed Norm, Duke Math. Jour., 28 (1961), 301-324.

[22] E. M. Stein, Interpolation of Linear Operators, Trans. Amer. Math. Soc., 83 (1956), 482-492.

MARTINGALE AND MAXIMAL THEOREMS

[23] D. G. Austin, A Sample Property of Martingales, Ann. Math. Statist., 37 (1966), 1396-1397.

[24] D. L. Burkholder, Martingale Transforms, *ibid.*, 1494-1504.

[25] D. L. Burkholder and Y. S. Chow, Iterates of Conditional Expectation Operators, Proc. Amer. Math. Soc., 12 (1961), 490-495.

[26] J. L. Doob, "Stochastic Processes," New York, 1952.

[27] _____, A Ratio Operator Limit Theorem, Zeit. Wahrscheinlich., 1 (1963), 288-294.

[28] R. F. Gundy, A Decomposition for L^1 Bounded Martingales, Ann. Math. Statist., 39 (1968), 134-138.

[29] R. E. Paley, A Proof of a Theorem on Averages, Proc. London Math. Soc., 31 (1930), 289-300.

[30] G. C. Rota, An "Alternieren de Verfahren" for General Positive Operators, Bull. Amer. Math. Soc., 68 (1962), 95-102.

[31] E. M. Stein, On the Maximal Ergodic Theorem, Proc. Nat. Acad. Sci., 47 (1961), 1894-1897.

ADDITIONAL REFERENCES

[32] S. Bochner, "Harmonic Analysis and the Theory of Probability," Univ. of Calif. Press, 1955.

[33] D. L. Burkholder, Semi-Gaussian Subspaces, Trans Amer. Math. Soc., 104 (1962), 123-131.

[34] J. Marcinkiewicz, Sur les multiplicateurs des séries de Fourier, Studia Math., 8 (1939), 78-91.

[35] F. Riesz and B. Sz-. Nagy, "Functional Analysis," New York, 1955.

[36] E. C. Titchmarsh, "Eigenfunction Expansions Associated with Second-order Differential Equations," Oxford, 1946.

[37] B. Muckenhoupt and E. M. Stein, Classical Expansions and Their Relation to Conjugate Harmonic Functions, Trans. Amer. Math. Soc., 118 (1965), 17-92.

[38] B. Muckenhoupt, Poisson Integrals for Hermite and Laguerre Expansions, Trans. Amer. Math. Soc., 139 (1969), 231-242; Hermite Conjugate Expansions, *ibid.*, 243-260.

[39] A. P. Calderón, Intermediate Spaces, and Interpolation, Studia Math., 24, (1964), 113-190.

[40] K. Yosida, "Functional Analysis," Berlin, 1965.

[41] G. Gasper, Jr., On the Littlewood-Paley and Lusin Functions in Higher Dimensions, Proc. Nat. Acad. Sci., 57 (1967), 25-28.

APPENDIX (1985)

In the fifteen years since the publication of this monograph, considerable progress has been made in several areas treated here. We indicate briefly some of this work, giving also the principal references.

1. More thorough going "probabilistic" approaches to the material of Chapters III and IV (leading to some further results) have been developed by P. A. Meyer [5], [6], and N. Varopoulus [11].

2. A different way to obtain Corollary 3 in Chapter IV, which proceeds by the method of "transference", is described in Coifman and G. Weiss [2], §4.

3. A wider class of multipliers applicable in the setting of the general theory and other related results is presented in Cowling [3].

4. The motion of sub-harmonicity (as used in Chapters II and V) has been generalized by Bakry [1], where several applications are given. Further results in the Sturm-Liouville case are also in Rachdi's thesis [7].

5. A vector-valued version of the maximal theorem (answering a question in Chapter IV, §3) can be found in Fefferman and Stein [4].

6. A more extended treatment of L^p multipliers of the Marcinkiewicz type on compact Lie groups is given by N. Weiss [12]. An alternative approach based on a functional calculus for pseudo-differential operators was carried out by Strichartz [10].

7. The ideas of this monograph can be used to deal with questions of harmonic analysis in $\mathbf{R}^n$, as $n \to \infty$; see [5], [8], and Stein and Strömberg [9].

REFERENCES

[1] D. Bakry, "Transformations de Riesz pour certaines semi-groupes symétriques," C.R. Acad. Sci. Paris 299 (1984), 999-1000.

[2] R.R. Coifman and G. Weiss, "Transference methods in analysis," Regional Conference Series in Mathematics 31 (1977), Amer. Math. Soc.

[3] M. Cowling, "Harmonic analysis on semi-groups," Ann. of Math. 117 (1983), 267-283.

[4] C. Fefferman and E.M. Stein, "Some maximal inequalities," Amer. J. Math. 93 (1971), 107-115.

[5] P.A. Meyer, "Démonstration probabiliste de certaines inéqalités de Littlewood-Paley," Sém. Prob. X, pp. 125-183 in Lecture Notes in Math. 511 (1976) Springer; Corrig. in Sém. Prob. XV, 850 (1981).

[6] ———, "Transformations de Riesz pour les lois gaussiennes," Sém. Prob. XVIII in Lecture Notes in Math. 1059 (1984) Springer.

[7] L.T. Rachdi, "Conjecture de Stein pour des opérateurs differentials singuliers sur un intervalle borne," Thesis, University of Tunis, 1984.

[8] E.M. Stein, "Some results in harmonic analysis in $\mathbf{R}^n$, for $n \to \infty$," Bull. Amer. Math. Soc. 9 (1983), 71-73.

[9] E.M. Stein and J.O. Strömberg, "Behavior of maximal functions in $\mathbf{R}^n$ for large n," Arkiv f. Mat. 21 (1983), 259-269.

[10] R.S. Strichartz, "A functional calculus for elliptic pseudo-differential operators," Amer. J. Math. 94 (1972), 711-722.

[11] N. Varopoulos, "Aspects of probabilistic Littlewood-Paley theory," J. Functional Analysis 38 (1980), 25-60.

[12] N. J. Weiss, "L^p estimates for bi-invariant operators on compact Lie groups," Amer. J. Math. 94 (1972), 103-118.

Lightning Source UK Ltd.
Milton Keynes UK
UKHW012229080922
408557UK00001B/164

9 780691 080673